T0073745

# Inside the UN Security Council

# Inside the UN Security Council

*Legitimation Practices and Darfur*

JESS GIFKINS

OXFORD
UNIVERSITY PRESS

# OXFORD
## UNIVERSITY PRESS

Great Clarendon Street, Oxford, OX2 6DP,
United Kingdom

Oxford University Press is a department of the University of Oxford.
It furthers the University's objective of excellence in research, scholarship,
and education by publishing worldwide. Oxford is a registered trade mark of
Oxford University Press in the UK and in certain other countries

Published in the United States of America by Oxford University Press
198 Madison Avenue, New York, NY 10016, United States of America

British Library Cataloguing in Publication Data

Data available

Library of Congress Control Number: 2023949047

ISBN 9780192869029

DOI: 10.1093/oso/9780192869029.001.0001

Printed and bound by
CPI Group (UK) Ltd, Croydon, CR0 4YY

MIX
Paper | Supporting
responsible forestry
FSC® C013604

*For Roger and Jenny Gifkins*

*and*

*Izzy Jayasinghe*

# Acknowledgements

This book began as PhD research within the Asia-Pacific Centre for the Responsibility to Protect at the University of Queensland, and I am endlessly grateful to the excellent guidance of Alex Bellamy and Phil Orchard as PhD supervisors and beyond. The PhD-to-book process has taken place over a long period of time, and I have benefited immensely from other projects alongside the revisions, especially multiple projects with Jason Ralph, whose influence is throughout this book. Concurrent projects with Stephen McLoughlin, Victoria Basham, Sam Jarvis, Dean Cooper-Cunningham, and the excellent staff at Protection Approaches have also inspired and sustained me.

I have benefited immensely from colleagues and friends reading chapters, some multiple chapters, for which I am grateful. This includes Alex Bellamy, Phil Orchard, James Pattison, Jason Ralph, Stephen McLoughlin, Tim Aistrope, Victoria Basham, Mark Kersten, Alex Beresford, Roger Mac Ginty, Daniel Wand, Thomas Dörfler, Georgina Waylen, Ben Walter, Kim Nackers, and Nora Gotzmann.

I presented earlier versions of this project at workshops, seminars, and conferences and received generous feedback, especially from the Critical Global Politics cluster seminar series at the University of Manchester, the University of Copenhagen Politics department, and conferences at ISA, BISA, EISA, and ACUNS.

I am grateful to my colleagues at the University of Manchester Politics department and have been lucky to have the opportunity to create two modules on the United Nations, including one on practices, and to discuss ideas with students. I am grateful to Dominic Byatt as Commissioning Editor with Oxford University Press for his support throughout and the three anonymous reviewers who were generous and engaged.

I did my PhD with a great cohort at the Asia-Pacific Centre for the Responsibility to Protect, especially Alex Pound, Stephen McLoughlin, Sarah Teitt, Charlie Hunt, Luke Glanville, Kim Nackers, Raymond Lau, and Eglantine Staunton. The Feminist Reading Group at the University of Queensland made the process of writing a PhD all the richer and more fun. I now have the privilege to supervise, and learn from, excellent PhD researcher Qiaochu Zhang.

I moved from Australia to the UK in 2013, and the ESRC Seminar Series on the Responsibility to Protect, led by Jason Ralph, James Pattison, Adrian Gallagher, and Aidan Hehir was an invaluable source of networking and integration into the

UK R2P community. I am grateful for support and friendship over the years from the European Centre for the Responsibility to Protect and the Global Centre for the Responsibility to Protect, especially Cristina Stefan and Benedict Docherty.

I was fortunate to conduct research interviews for this project and am grateful to all the interviewees who gave up their time including Jean-Marie Guéhenno, Colin Keating, and Edward Luck, and others who chose to be anonymized. Loraine Sievers has patiently answered obscure questions on Security Council procedures. Paul Williams and Richard Gowan offered generous insights into the UN over many years. I am grateful to James Thomson who introduced me to the responsibility to protect during a Masters internship in 2007 and Aguil Deng who encouraged me to research Sudan, all those years ago. An earlier version of chapter one was published in *Global Governance* as Jess Gifkins (2021) 'Beyond the Veto: Roles in UN Security Council Decision-Making', *Global Governance* 27(1): 1–24 and I am grateful to the publishers for permission to use this material.

Friends including Rebecca Shaw, Stephen McLoughlin, Elaine Kelly, Andrea Gaion, and Laura Parmisciano have supported and encouraged me throughout. I am grateful for endless support from both the Gifkins and Jayasinghe families.

I owe my biggest debt of gratitude to Izzy Jayasinghe, a brilliant scholar in her own right, who has always believed in me, supported me, and encouraged me.

# Contents

# Abbreviations

| | |
|---|---|
| AFP | Agence France-Presse |
| AMIS | African Union Mission in Sudan |
| AP | Associated Press |
| AP/ST | Associated Press/Sudan Tribune |
| AU | African Union |
| BBC | British Broadcasting Corporation |
| CRED | Centre for Research on the Epidemiology of Disasters (Belgium) |
| CPA | Comprehensive Peace Agreement |
| DFID | Department for International Development (UK) |
| DPA | Darfur Peace Agreement |
| DPKO | Department of Peacekeeping Operations (UN) now the Department of Peace Operations |
| DPO | Department of Peace Operations (UN) |
| E10 | Elected Ten members of the UN Security Council |
| G20 | Group of Twenty |
| GAO | Government Accountability Office (US) |
| HRW | Human Rights Watch |
| ICC | International Criminal Court |
| ICG | International Crisis Group |
| ICID | International Commission of Inquiry on Darfur |
| ICISS | International Commission on Intervention and State Sovereignty |
| JEM | Justice and Equality Movement |
| MSF | Médecins Sans Frontières |
| NATO | North Atlantic Treaty Organization |
| NGO | Non-Governmental Organization |
| OCHA | Office for the Coordination of Humanitarian Affairs (UN) |
| P3 | Western permanent three members of the UN Security Council |
| P5 | Permanent five members of the UN Security Council |
| PoC | Protection of Civilians |
| QRM | Quick Review Mission |
| R2P | The Responsibility to Protect |
| SLM/A | Sudan Liberation Movement/Sudan Liberation Army |
| ST/AFP | Sudan Tribune/Agence France-Presse |
| TAM | Technical Assessment Mission |
| TCC | Troop Contributing Country |
| UK | United Kingdom of Great Britain and Northern Ireland |
| UN | United Nations |
| UNAMID | United Nations/African Union Hybrid Operation in Darfur |

| | |
|---|---|
| UN DPA | United Nations Department of Political Affairs[1] |
| UNHCR | The Office of the United Nations High Commissioner for Refugees |
| UNICEF | United Nations Children's Fund |
| UNMIS | United Nations Mission in Sudan |
| US | United States of America |
| USAID | United States Agency for International Development |
| USSR | Union of Soviet Socialist Republics |

[1] I use UN DPA over the standard acronym DPA because I use DPA to refer to the Darfur Peace Agreement.

# Introduction

*Few institutions are as well-known or as little understood as the UN Security Council*

(Luck 2006: 127)

The UN Security Council wields unprecedented power in global governance, with the primary responsibility to maintain international peace and security. Its decisions affect the lives of billions of people, yet its day-to-day decision-making takes place in informal settings with little public record or scrutiny. The formal rules governing this body—as outlined in the UN Charter and the Provisional Rules of Procedure—provide a 'skeleton' for understanding membership and voting in the Security Council but tell us little about how it operates on a day-to-day basis (Schia 2013: 140). A skeleton gives important information about the structure and stability of a body but tells us little about how it functions or sustains itself. Similarly, the Security Council's formal rules offer little insight into decision-making. Instead, informal practices shape, structure, and define negotiations. While resolutions pass or fail at public meetings at the iconic horseshoe table, the actual decision-making takes place in closed meetings and in a variety of informal settings. To explain informal negotiations this book introduces the concept of legitimation practices, which are practices that enhance the legitimacy of a decision, or those negotiating the decision, and are central to negotiations. Legitimation practices include, for example, prioritizing unanimity, repeating previously agreed language, and maintaining consistent arguments. While there is a significant body of literature on the legitimacy or illegitimacy of the Security Council and its decisions, research has only scratched the surface of the impact of legitimation on the process and outcome of negotiations. Informal legitimation practices are crucial to understanding how the Security Council makes decisions.

This book analyses patterned actions whereby states seek legitimacy for themselves as actors or for a Security Council decision, as part of Council negotiations. To analyse legitimation—the processual dimension of legitimacy—this book builds on rapidly growing interest within International Relations on practice theory (see for example Adler and Pouliot 2011a, 2011b; Brown 2012; Bueger and Gadinger 2018; Frost and Lechner 2016; Hopf 2018; Pouliot 2008; Ralph and Gifkins 2017; Schindller and Wille 2019). Practice theory refocuses analytic attention onto day-to-day interactions between actors, which International

*Inside the UN Security Council*. Jess Gifkins, Oxford University Press. © Jess Gifkins (2023).
DOI: 10.1093/oso/9780192869029.003.0001

Relations has historically tended to eschew in favour of state-centric approaches (Adler-Nissen 2015). For this reason, the practice turn within International Relations is particularly suitable for studying diplomacy and has been widely used in recent years (Adler-Nissen 2016; Adler-Nissen and Drieschova 2019; Cornut 2017; Kuus 2015; Neumann 2002; Pouliot and Cornut 2015; Wiseman 2015). Practices can be thought of as a 'theory-laden way to refer to what people do in everyday life' (Nicolini 2012: 54). As Adler and Pouliot have explained, 'Practices are competent performances. More precisely, practices are socially meaningful patterns of action, which, in being performed more or less competently, simultaneously embody, act out, and possibly reify background knowledge and discourse in and on the material world' (2011a: 4). Practices rely on audience judgements and are social by nature. As such, this book both supplements existing practice-based accounts of the Security Council and adds a novel legitimation dimension to existing explanations of practices in this institution (Adler-Nissen and Pouliot 2014; Bode 2018; Cook 2016; Niemann 2022; Pouliot 2021; Ralph and Gifkins 2017). Practice theory is well suited to this book because it is process-oriented rather than outcome-oriented, so is ideally suited to analyse the process of negotiation.

This book diverges from the standard approach of analysing legitimation of an institution or its decisions, to analysing the ways that legitimation practices shape the negotiation process and the resulting decisions. Using practice theory, the book analyses the way that considerations of legitimacy—by states—affect the process and outcome of Security Council negotiations. Interest has increased considerably in the relationship between legitimacy and the UN Security Council in the twenty-first century, however, the existing literature focuses on the legitimacy (or otherwise) of the Security Council as an institution and the legitimacy (or otherwise) of its decisions (Binder and Heupel 2014; Frederking and Patane 2017; Hurd 2002, 2007; Morris and Wheeler 2007; Welsh and Zaum 2013). There is also a growing body of literature on legitimation in relation to the United Nations, however, this literature focuses on assessing and explaining the legitimation of the UN as an institution rather than how legitimation impacts decision-making (von Billerbeck 2019, 2020; Welsh and Zaum 2013; Westra 2012; Zaum 2013). Shifting orientation to legitimation within negotiations, this book draws on the Security Council's response to the conflict in Darfur in the west of Sudan as a case study, which was one of the most significant humanitarian crises in the early twenty-first century.

The conflict in Darfur escalated in 2003 after the government of Sudan responded to military success of rebel groups with comprehensive violence against people from the same ethnic group as the rebels. This book analyses the decisions taken by the Security Council between 2004 and 2007, which includes decisions to apply sanctions, to refer the situation in Darfur to the International Criminal Court, and to authorize a peacekeeping operation. It does not include the conflict in Sudan and Darfur that escalated in 2023 due to publishing schedules.

The scale of human suffering in Darfur in the early 2000s was immense and it is widely recognized as a case where the international community failed to respond effectively to the crisis (Badescu and Bergholm 2009; Grünfeld, Vermeulen and Krommendijk 2014; Lanz 2020; Mills 2009; Prunier 2008). The contentious and divided nature of decisions on Darfur meant that the Security Council often negotiated decisions over months rather than days. This in turn means that there is considerable material available for researching the negotiations, such as newspaper articles, statements, memoirs, and reports, which enables greater access to the negotiation process than was often the case during this period for quicker negotiations.[1] The Security Council's engagement with Darfur is also typical of broader patterns, as an intrastate conflict, located in Africa, which is the region the Council is most active in (Deplano 2015; Dunbabin 2008). The Security Council also used escalating tools, known as ratcheting, which is common in its responses to other situations (Gehring and Dörfler 2019). By using a thick analysis of Security Council negotiations on Darfur, this book demonstrates informal practices across the remit of the Security Council's work.

This book introduces the concept of internal and external legitimation practices, building on the idea of legitimacy as a 'social status' that states seek, which is internal or external from the vantage point of the Security Council (Coleman 2007: 20). As such, 'internal' relates to the legitimacy of a decision and 'external' relates to the legitimacy of an actor. Internal legitimation practices are actions that drafters take during negotiations to enhance the legitimacy of a decision. Examples of this include seeking unanimity for a decision even though this is beyond the legal requirements for a resolution, seeking support from states who are in the region the decision relates to, and repeating language from a previous decision as a means of circumventing dissent. These internal legitimation practices are common, yet informal, and both bolster the legitimacy of a decision and mitigate against dissent.

External legitimation practices occur when an actor strives for legitimation from its key audiences and this shapes its actions in the Security Council. External legitimation practices include states changing positions because an argument they attempted to make was inconsistent with previous arguments they had made and generated negative attention. It also includes states choosing to take action, even if they know in advance that the decision cannot be implemented, so they look like they are 'doing something'. In each of these scenarios—consistency in arguments or 'doing something'—states take specific steps in Security Council negotiations to enhance their legitimacy with audiences that are important to them. External legitimation practices enable Security Council members to secure sufficient agreement

---

[1] Since 2011, there has been much greater access to the detail of negotiations via Security Council Report's 'What's in Blue' website, which publishes detail of negotiations in real time. For detail see https://www.securitycouncilreport.org/whatsinblue

to pass a decision and internal legitimation practices shape the resulting decisions. These practices are integral to Security Council negotiations and are currently an underresearched area.

Foregrounding legitimation practices sheds light on seemingly contradictory moments within Security Council decision-making. For example, former US Ambassador John Danforth led negotiations towards the first two resolutions on Darfur that threatened wider sanctions against the government of Sudan. The Security Council did not follow through on this threat, despite a lack of compliance from Khartoum, and the following year, Danforth explained what he understood at the time: 'I am 100% certain. There was no chance of getting sanctions' (Keane 2005). What was it about the process of decision-making that enabled these two resolutions to pass—including a threat of sanctions—despite a frank awareness by those leading the negotiations that it was not possible to implement sanctions? This example is explained in Chapter 5 in terms of the legitimation benefits for states of 'doing something', or at least, appearing to 'do something'.

Similarly, the US was fundamentally opposed to referring the situation in Darfur to the International Criminal Court (ICC) and had the power to block this decision via a negative vote. Rather than veto, however, the US abstained from the vote but stated bluntly, 'we do not agree to a Security Council referral of the situation in Darfur to the ICC' (United Nations 2005: 3). The US not only did not agree with the referral but also had fundamental concerns about the ICC as an institution. The US Ambassador continued: 'we have not dropped, and indeed continue to maintain, our long-standing and firm objections and concerns regarding the ICC' (United Nations 2005: 3). Yet, at the final hour, the US chose not to block this referral and abstained from the vote instead. There was a legitimation value for the US in maintaining a consistent argument on Darfur, especially since it has established itself as a key advocate on Darfur domestically and in the Security Council, as is set out in Chapter 6. Analysing legitimation practices within negotiations is useful for explaining these seemingly contradictory key moments in decision-making.

Legitimation practices in the UN Security Council are worth analysing for three key reasons. First, the Security Council wields unprecedented power within global governance. It has the capacity to make binding decisions on all UN members, commands around a hundred thousand peacekeepers in about a dozen countries, and has the power to refer situations to the ICC, which, in turn investigates individuals on charges of war crimes, crimes against humanity, genocide, and the crime of aggression. This means that Security Council decisions have a significant impact on global politics and that understanding how these decisions are made is paramount. Second, even experienced diplomats highlight the steep learning curve they face upon taking up a seat within the UN Security Council, and the ways that their perspective on the institution changes (Rosenthal 2017). This suggests that decision-making in the Security Council is unique, even when compared to other forms of multilateral diplomacy. Third, there is widespread acknowledgement of the legitimation role of the Security Council and the centrality of legitimacy for the implementation of its decisions (Claude 1966; Hurd 1999, 2002, 2007; Thompson 2006, 2010). Yet there is not yet a commensurate literature on how legitimacy functions within the process of decision-making. Documenting and analysing informal practices of the UN Security Council is a worthwhile venture, then, because of the significance and uniqueness of the institution, and the limited literature on legitimacy in the process of reaching decisions.

As recent debate has shown, International Relations as a discipline often overlooks the minutiae of daily exchanges through which international relations is enacted (Adler-Nissen 2015; Adler-Nissen and Drieschova 2019; Adler-Nissen and Pouliot 2014; Wiseman 2015). This lack of focus on ideational factors is congruent with rationalist assumptions that are often made about the nature of international decision-making: that interests are exogenous, and that the negotiations inside international organizations are a reflection of power outside the institution (Glennon 2003; Mearsheimer 1994/1995; Thompson 2010). Yet many 'insider' accounts of the UN Security Council stress the influence of informal practices in shaping decisions, even if they do not necessarily use the language of 'practices' to explain their experiences (Barnett 1997; Fasulo 2015; Rosenthal 2017; Schia 2013, 2017; Smith 2006). Practices, and more specifically, legitimation practices, are central to the negotiation of Security Council decisions, and analysing these enables a richer account of negotiations.

The past two decades have seen a growing interest in the relationship between legitimacy and the UN Security Council. It is unsurprising that there has been an explosion of literature on the legitimacy of the Security Council in the post–Cold War period, given the expansion of its mandate, and high-profile failures of the Security Council such as the Rwandan genocide and the use of safe havens in Bosnia and Herzegovina. As Frost highlights, 'When an institution is working well from the perspective of outside participants questions of legitimacy are rarely asked' (2013: 27). Indeed, according to Hurd, 'There is near constant agreement that the legitimacy of the Council is under threat' (2008: 199). Missing from this analysis, however, is a focus on how legitimacy shapes day-to-day practices within negotiations within the UN Security Council.

Although possessing the legal authority to make decisions that are binding on all UN members, the Security Council has limited capacity to enforce its decisions, meaning that the enactment of its decisions necessarily relies on the legitimacy bestowed on the institution. Given this, the growing literature on legitimacy and the Security Council has tended to focus on questions relating to the legitimacy of the institution (Buchanan and Keohane 2011; Hurd 2007, 2008; Welsh and Zaum 2013), why powerful states choose to act through the Security Council (Chapman and Reiter 2004; Thompson 2006, 2010), and how the Security Council is perceived (Binder and Heupel 2014). This literature provides useful insights into the normative standing of the Security Council and the relationships between states and the Council. It shows why states want to be part of the Security Council and why they want the Council to address specific situations (for either symbolic or signalling reasons). Once states are in the Security Council and situations are under consideration by the Council, it does little, however, to explain how legitimacy then shapes the process of negotiation.

Rather than assessing the legitimacy of the Security Council as an institution, this book analyses the way that considerations of legitimacy shape the process and outcome of decision-making. Standard definitions of legitimacy within International Relations come from a democratic model whereby an actor feels compelled to comply with a rule or an institution (see for example Franck 1990: 16; Hurd 2007: 7). These definitions are congruent with a constructivist analysis demonstrating the 'compliance pull' that norms have, to use Franck's language (1990). The use of legitimation practices in this book necessitates a more social and relational understanding of legitimacy because the focus is not on states complying with rules but on states seeking the legitimation that comes from taking actions that appeal to the audiences that are important to them. As such, this book draws on a social and relational version of legitimacy, congruent with that put forward by Coleman, where legitimacy can be conceptualized as a 'social status that can adhere to an actor or an action' (2007: 20). This broader understanding of legitimacy encompasses aspects of decision-making that a focus on institutions and rules does not. Understandings of legitimacy vary, so the types of actions and behaviours that have a social status of legitimacy will be contingent on a particular group or institution at a particular time (Hurrell 2005). Focusing on considerations of legitimacy in the decision-making process, this book adopts a more social understanding of legitimacy.

## Gaps in the Existing Literature

Rationalism is a common starting point in literature on the UN Security Council, particularly focused on why states use the Security Council, rather than how the Council makes decisions. Examples of this include, first, research on the signalling implications of acting through the Security Council and why powerful states might choose to act through the Security Council (Chapman and Reiter 2004; Gruber 2005; Matsumura and Tago 2019; Thompson 2006, 2010; Voeten 2001, 2005). This literature shows that states may make (albeit slight) concessions to gain the signalling advantages that come from Security Council approval (see for example Thompson 2010: 25–26). Second, research on the changes in foreign aid received by elected members from the US, UN funds and agencies, including the International Monetary Fund, and the World Bank during their term on the Council (Bashir and Lim 2013; Bueno de Mesquita and Smith 2010; Dreher,

Sturm and Vreeland 2006, 2009, 2015; Kuziemko and Werker 2006). This literature suggests that powerful states incentivize the support of elected members by
granting additional foreign aid or aid with less conditionalities, however, there
is limited evidence of this directly influencing the votes of elected members (for
example see Eldar 2008). Third, there is a group of literature analysing which situations the Security Council responds to and how these cases relate to interests of
permanent members of the Council (Allen and Yuen 2013; Beardsley and Schmidt
2012; Binder 2015). These three groups of literature offer some insights into why
states seek to use the Security Council, how states leverage aid to encourage support, and which situations are (and are not) considered by the Security Council.
Approaches to the above research questions typically assume that states are ratio-
nal actors, and there is more emphasis on the outcome of negotiations than the
process of negotiations.

The assumption of fixed exogenous state interests is also a feature of explanations of Security Council decision-making in specific cases. Assumptions about
exogenous interests, particularly material interests, on the part of permanent five
(P5) members are inherent in explanations of Security Council decision-making
on Darfur. There are three main interest-based explanations for the Security
Council's action—and lack thereof—on Darfur. First, China's oil interests in Sudan
are often cited to explain the behaviour of China (Adelman 2010; Badescu and
Bergholm 2009; Belloni 2006; Caplan 2006; Grünfeld, Vermeulen and Krommendijk 2014; Prunier 2008; Straus 2006). Prunier provides a stark example of
this line of argument, although many other authors share this sentiment: 'China
holds a large share of responsibility in the ongoing Darfur horror. The reason is
exceedingly simple: oil' (2008: 178). Second, Russia's material interests in selling
arms to Sudan and purchasing Sudanese oil are regularly used to explain Russia's behaviour on Darfur (Badescu and Bergholm 2009; Belloni 2006; Grünfeld,
Vermeulen and Krommendijk 2014). Third, the United States' interests in coop-
erating with Sudan for intelligence information are often used to explain the US's
actions in negotiating on Darfur in the Security Council (Badescu and Bergholm
2009; Caplan 2006; Grünfeld, Vermeulen and Krommendijk 2014; Stedjan and
Thomas-Jensen 2010). These material interest-based explanations are incomplete
rather than incorrect.

These rationalist arguments go part of the way towards explaining the behaviour
of states in these negotiations; however, an analysis of legitimation practices can
explain changes in positions that enabled the Security Council to reach decisions on Darfur, where earlier material interest arguments cannot. For example,
the material interest-based accounts do not explain why the United States shifted
position from resisting any Security Council action on Darfur to actively championing it. They do not explain why the US enabled the Security Council to refer
the situation in Darfur to the ICC, despite its serious concerns with the ICC as
an institution. They also do not explain why China's position on the primacy of

consent for peacekeeping evolved in the case of Darfur to include pressure from Beijing to induce consent from Khartoum. To explain these changes in positions we need to go inside the Security Council and examine legitimation practices in decision-making.

## Outline of the Argument

This book introduces and develops the concept of legitimation practices. It argues that these were critical in enabling and shaping the UN Security Council's decisions on Darfur and have wider relevance for understanding Council negotiations beyond this case. There is also scope to apply legitimation practices to other multilateral settings or to domestic governance. Legitimation practices are repeated patterns of behaviour that serve to enhance or reinforce the legitimacy of an actor or a decision. While this concept would have broader utility beyond the Security Council, for the purposes of this book legitimation practices refer to actions taken to enhance the legitimacy of a state involved in the negotiations or a Security Council resolution. There has been considerable interest in legitimacy in relation to the Security Council over the past two decades, however, this has typically focused on the legitimacy (or otherwise) of the Security Council as an institution or the legitimacy (or otherwise) of Security Council decisions. This book instead analyses legitimation within the decision-making process. The book focuses on the process of negotiation, and therefore benefits from drawing on the recent growth of interest in practice theory within International Relations. Framing legitimation as a practice highlights the repetitive nature of legitimation activities and the ways in which these are meaningful to participants, drawing on the definition of practice used by Adler and Pouliot (2011a, 2011b). There are a small number of recent articles which refer to legitimation practices, however, following general trends, the focus of these is on the legitimation of the international organization itself, rather than legitimation within negotiations (Spandler 2020; Tallberg and Zürn 2019; von Billerbeck 2019, 2020). This book instead develops the concept of legitimation practices in relation to negotiations.

This book develops the concepts of internal and external legitimation practices. External legitimation practices—whereby an actor seeks legitimacy in the eyes of its key audiences and this shapes its actions in the Security Council—were instrumental in reaching decisions. The two external legitimation practices identified are maintaining consistency in arguments, and the incentivized nature of 'doing something', or at least being seen to be doing something by taking actions even if they cannot be implemented. Across each of the issue areas considered in this book—setting the agenda, sanctions, referral to the International Criminal Court, and peacekeeping—permanent members were initially strongly opposed to taking action and in each instance it was external legitimation practices that changed the position of the states in question and enabled a decision to be reached. For example, positioning itself as a vocal advocate on Darfur subsequently made it difficult for the US to veto a referral of the situation to the ICC, even though it was fiercely critical of the ICC. Maintaining a consistent argument on Darfur incentivized the US to shift position and enable the ICC referral, despite considerable concerns about the referral.

Alongside external legitimation practice shaping key turning points, there are also internal legitimation practices identified in this book that shaped decisions once they were under negotiation in the Security Council. Internal legitimation practices—actions taken by drafters to enhance the legitimacy of a decision—were instrumental in shaping the contours of decisions, and at times, were instrumental in reaching decisions at all. The three internal legitimation practices identified in this book are seeking unanimity, which is beyond the legal requirements, seeking support from states in the region under discussion, and drawing on prior language as a means of smoothing negotiations. These internal legitimation practices are not part of the formal rules but have evolved informally as a shared understanding between Security Council members, particularly permanent members who have more capacity to shape daily practices over the long term. The goal here is to explain the role of legitimation practices in enabling and shaping decision-making in the UN Security Council. Internal and external legitimation practices are key in deepening the understanding of decision-making in the Security Council.

Actors within the Security Council have different capacity to influence negotiations, yet there is limited literature on the different types of influence that actors have. For example, it has been said that although there is awareness that permanent members dominate negotiations—and particularly Western permanent members—there is limited understanding of how this dominance is enacted (Adler-Nissen and Pouliot 2014: 909). This book uses the concept of institutional power, as set out by Barnett and Duvall, to explain *how* permanent members dominate negotiations, above and beyond their Charter-given privileges (Barnett and Duvall 2005a, 2005b). The UN Charter gives permanent members ongoing tenure and the capacity to block decisions (known colloquially as a 'veto', although this term does not appear in the Charter). However, in practice, permanent members dominate all aspects of negotiations, from agenda-setting, to drafting, to their capacity for informal veto. Set against this backdrop, some have argued that elected members have close to zero capacity to influence negotiations (Junn 1983; O'Neill 1996). There are ways that elected members can influence negotiations though, which draw from internal legitimation practices in the Security Council. Drafters—who tend to be permanent members—prioritize the legitimacy that comes from unanimity and they prize support from states located in the same region as the conflict under discussion. Elected members can also magnify their influence via acting collectively or via the diplomatic skill of their permanent mission. Although the UN Secretariat is not a voting member of the Security Council, it too can have influence over decisions either by shaping the normative environment of decision-making. The other group of actors which shape advocacy and framing on issues under discussion in the Security Council are the epistemic community around the UN, also known as the 'third UN', including NGOs, academics, and media organizations. Institutional power is useful for explaining how permanent members extend their influence beyond their formal privileges, yet internal legitimation practices mitigate the extent of this influence. The Secretariat and third UN speak from positions of legitimated authority, which can also shape negotiations. As such, the institutional power of permanent members is not absolute and internal legitimation practices mitigate this influence.

In addition to developing the theoretical concepts of legitimation practices, this book also makes an empirical contribution to the literature on Security Council negotiations on Darfur. Analysing the Security Council's response to the situation in Darfur through the lens of legitimation practices reveals a more nuanced account of state behaviour. A common representation of state behaviour in relation to Darfur is of Western—particularly US—advocacy against Chinese opposition. Although cognizant of the delays on the part of Western states, Slim describes that from April 2004 'there was obviously significant and powerful political will, led by the United States itself, to stop the suffering in Darfur' (2004: 826). Similarly, Human Rights Watch describe the US as state that 'has taken the strongest public stance on Darfur' in May 2004 (2004: 58). China was described by Human Rights Watch as 'the main impediment to stronger action by the Security Council' (Clough 2005: 7), a sentiment echoed by Straus (2006: 194) and Prunier (2008: 178). Yet a detailed analysis of negotiations as set out in the empirical chapters shows that the US the UK were actually the primary opponents during the early stages of agenda-setting, with some US opposition in relation to targeted sanctions, and strong US opposition on referring the situation in Darfur to the ICC. China's actions also do not fit into a simple binary narrative of opposition opponent. Beijing resisted aspects of the negotiations on both sanctions and peacekeeping, but there are also examples where China was instrumental in encouraging affirmative votes from both Russia and Qatar for a resolution on sanctions and peacekeeping.[2] Similarly, after repeatedly blocking a decision in the Sanctions Committee, China allowed the same decision to pass the in the Security Council as it did not want to cast an isolated veto.[3] China was also instrumental in eventually securing Sudanese consent for hybrid peacekeeping between the United Nations and the African Union.[4] Material and strategic interests are part of the explanation for the actions of the US and China in negotiations on Darfur, but they do not offer a complete explanation. An account based on legitimation practices provides

---

[2] Resolution 1679. See Chapter 7 for discussion.
[3] Resolution 1672. See Chapter 5 for discussion.
[4] See Chapter 7 for discussion.

explanations for changes in state behaviour—from resisting an action to enabling an action—and explains key turning points in negotiations on Darfur.

This book is not normative in focus; however, it does come from a position of wanting Security Council decisions to be more effective for both the protection of civilians (PoC) and for the responsibility to protect (R2P). As Bode and Karlsrud have argued, while there is widespread support for the goals of PoC there is still considerable variation in implementation (2019). Similarly, broad rhetorical support for R2P has not translated into effective responses in many cases (Bellamy and Luck 2018; Gözen Ercan 2022; Welsh 2016). The focus in this book is on the process of reaching decisions, rather than the implementation of these decisions, so implementation is only specifically analysed when it feeds back into subsequent decisions. What does become apparent, however, is that some of the legitimation practices identified in this book, especially the practice of 'doing something', can be detrimental to meaningful action on the ground because it involves taking a decision to give the appearance of action, even when there is prior knowledge that a decision cannot be implemented. In relation to Darfur, there were extensive negotiations on both sanctions and peacekeeping, spanning months and using significant diplomatic capital, which were not implemented. As my co-author and I have argued elsewhere, there is a need for permanent members of the Security Council to use R2P-compentent practices, 'harmonizing human protection and collective action' (Ralph and Gifkins 2017: 645). The legitimation practices identified in this book pose some challenges for this form of collective action.

## The Conflict in Darfur as a Case Study

When writing the forward for Malone's monograph on the UN Security Council's decision-making on Haiti in 1998, Adam Roberts lamented, 'We need more such studies of how the UN Security Council actually works in crises' (Malone 1998: xii). The intervening two decades have seen few further monographs documenting the UN's decision-making in relation to specific cases.[5] This monograph analyses Security Council decision-making in response to the conflict in Darfur. There were aspects of the Security Council's engagement with Darfur that were unique. For example, after initially resisting discussion on Darfur, the level of engagement by the Security Council and the broader UN system was substantial. Indeed, Darfur became such a key part of the United Nations' work that by 2006 almost ten per cent of the UN Security Council's resolutions were on Darfur (Gifkins 2016b). The conflict in Darfur also invoked a range of national interests for permanent members, which both enabled and constrained action, and it became a cause célèbre in some Western capitals (Lanz 2020). The United Nations'

---

[5] Notable exceptions include Malone (2006) and Thompson (2010).

response to Darfur also involved aspects of innovation: an extensive partnership with the African Union on joint peacekeeping, the first referral of a case to the ICC by the Security Council, and one of the first cases that was debated in terms of the international community's 'responsibility to protect' (for discussion on Darfur and R2P see Gifkins 2016a). The scale of the Security Council's response to Darfur, along with the innovations made in this case, make it a case that warrants further analysis. Darfur was also one of the most contentious cases under discussion in the Security Council at the time. As Figure 1 shows, the vast majority of drafts voted on in the Security Council between 2004 and 2007 not only passed but also passed unanimously. As Figure 2 shows however, once drafts on Darfur are isolated from the total numbers, it is evident that most decisions on Darfur were not unanimous. As such, Darfur was one of the most contentious cases under negotiation at the time.

There are also aspects of the Security Council's engagement with Darfur that are representative of its wider patterns of decision-making on other cases. The Security Council is more engaged with conflicts in Africa than any other region,

| Year | Resolutions voted on | Resolutions Failed | | Resolutions Passed | |
|---|---|---|---|---|---|
| | | Vetoed | Did not reach nine votes | Unanimous | Not unanimous |
| 2004 | 62 | 3 | 0 | 55 | 4 |
| 2005 | 71 | 0 | 0 | 68 | 3 |
| 2006 | 89 | 2 | 0 | 83 | 4 |
| 2007 | 57 | 1 | 0 | 52 | 4 |

**Figure 1** Security Council Resolutions 2004–2007 by Voting Outcome

*Note:* Data compiled by the author from UN (2022).

| Year | Resolutions Failed | Resolutions Passed | |
|---|---|---|---|
| | | Unanimous | Not unanimous |
| 2004 | 0 | 0 | 2 |
| 2005 | 0 | 0 | 2 |
| 2006 | 0 | 1 | 2 |
| 2007 | 0 | 1 | 0 |

**Figure 2** Key Security Council Resolutions on Darfur 2004–2007 by Voting Outcome

*Note:* Data compiled by the author from UN (2022).

with 43 per cent of resolutions between 1945 and 2014 addressing situations in African states (Deplano 2015: 6). When considering UN sanctions alone, this rises as high as 68 per cent of UN sanctions were imposed in Africa between 1991 and 2013 (Charron and Portella 2015: 1371). As a conflict in an African state, which included UN sanctions, the responses to the conflict in Darfur fit within common patterns of the Security Council's work. This is also true as a predominately intrastate conflict, with the vast majority of wars now occurring within states (Dunbabin 2008). Since the end of the Cold War., the Security Council has an increased focus on responding to intrastate wars and humanitarian crises (Bellamy 2016; Yamashita 2007). The Security Council's response to the crisis in Darfur follows a pattern of escalating measures (ratcheting) which is also typical in its responses (Gehring and Dörfler 2019). In the case of Darfur, the measures applied went through stages of agenda-setting, sanctions, referral to the ICC and, eventually, peacekeeping, which form the thematic content of the empirical chapters. In addition to these points of both uniqueness and representativeness, there is a further pragmatic reason for choosing Darfur as the focus of a study on decision-making. Many Security Council resolutions are negotiated quickly, over twenty-four hours or less, and it is difficult to find the depth of material needed for a detailed study on the process of reaching a decision in a negotiation that is so fast (Rosenthal 2017: 104). Since Darfur was a relatively contentious case for Security Council members, many decisions took six months to negotiate, making it a rich case study for the process of decision-making.

## Methodology

Practice theory is ideally suited to research on the processual dimension of negotiations and draws our attention to the 'how' of negotiations. It does, however, require access to detailed information on meetings, roles, daily schedules, and socialization. Security Council decision-making is challenging to research as it takes place in informal settings largely without public record, making it difficult to access information.[6] Wood argues, 'There is rarely any direct evidence of the drafting process on the public record, though occasionally the records of Security Council meetings may shed important light on individual drafting points' (2016: 13). However, the 'glare of publicity' does occur even in informal meetings (Johnstone 2003: 463). The formation of the think tank Security Council Report in 2004 has made a considerable impact on the transparency of the institution, especially since 2011, when its 'What's in Blue' website began daily reports documenting

---

[6] Tzanakopoulos argues that the Security Council has an ancillary legal requirement to be transparent (2014).

Security Council negotiations in real time.[7] Also the book and companion website for Sievers and Daw's 'The Procedure of the UN Security Council' has also helped to situate current and evolving practices within a historical context (2014). These innovations over the past fifteen years have helped to increase the transparency of Security Council negotiations.

Analysing historical case studies, however, remains challenging. Research for this book involved seeking information from many sources—starting with secondary academic literature and public sources and moving into less easily accessible material. I began with academic literature and then formal UN documents from inside and outside the Security Council: resolutions, meeting records, statements, and press releases. I then supplemented these sources with research from archives of newspapers and news wires—via the Factiva database—that contained excerpts from drafts, positions adopted by states, and obstacles confronted in the process of reaching decisions, which enabled me to construct a timeline leading up to each resolution. For each of the key decisions on Darfur, I read all newspaper articles that contained 'Darfur' and the relevant tool (sanctions/the International Criminal Court/peacekeeping) in the lead paragraph of the article for the six months before each resolution passed, focusing on practice and process, which was many hundreds of articles per decision. This news archive research facilitated as rich and detailed an analysis of the negotiation process as possible, given the limited documentation of Security Council negotiations.

To supplement UN and news sources, I also wanted to access perspectives from people who were directly involved in negotiations on Darfur. Given that Darfur was a high-profile case, some of these are now available on the public record in the form of memoirs by participants—including Kofi Annan (2012), Jan Egeland (2008), John Bolton (2007), and Jean-Marie Guéhenno (2015)—and media interviews with key diplomats. WikiLeaks released many cables, which gave detailed descriptions of informal meetings, briefings, and negotiations, albeit from a US perspective.[8] The cables offered rare access to some of the minutiae of decision-making, recorded as it unfolded, including records of conversations and positions. I supplemented these public sources with a series of elite semi-structured interviews with individuals who had direct involvement with key decisions on Darfur from UN Permanent Missions of Security Council members—including

---

[7] Security Council Report was formed in late 2004 and reported on the later sanctions negotiations on Darfur as well as the ICC referral, and debates on peacekeeping. From 2011, Security Council Report began a companion website called 'What's in Blue', named after the convention of printing Security Council decisions in blue ink during the final stages of negotiation. What's in Blue is an invaluable resource for analysing Security Council negotiations since 2011.

[8] The cables used in this book, released by Wikileaks, were primarily written by the US Permanent Mission in New York. This means they are necessarily written from a US perspective, for a US diplomatic audience, which is taken into consideration when evaluating the strength of sources. The cables have two key strengths: they were documented in 'real time' rather than after the fact; and they offer unusual access to Security Council negotiations, including private conversations with many fellow Security Council members and members of the UN Secretariat.

staff at the levels of ambassadors, political coordinators, and experts—and from departments of the UN Secretariat. I anonymized most of these interviews, at the request of interviewees, using a description of their choice, although Jean-Marie Guéhenno, Colin Keating, and Professor Edward Luck kindly allowed me to identify their material by name. Additionally, I conducted research interviews with individuals from think tanks and NGOs which were involved with the negotiations on Darfur at the time. These interviews focused on the process of negotiations.[9] This detailed empirical research—drawn from both public and private sources—enabled me to reconstruct the Security Council's negotiations on Darfur in the months leading up to the adoption of key resolutions and to triangulate information between sources.

My principal focus in the case study material is the process of negotiation. Practice theory offers the most useful analytical approach because it enables a focus on the 'how' of negotiations. As Pouliot explains, 'practice theory is a fundamentally abductive methodology based on the jotting together of empirics and analytics' (2016). Friedrichs and Kratochwil (2009) elaborate:

> The typical situation for abduction is when we, as social scientists, become aware of a certain class of phenomena that interests us for some reason, but for which we lack applicable theories. We simply trust, although we do not know for certain, that the observed class of phenomena is not random. We therefore start collecting pertinent observations and, at the same time, applying concepts from existing fields of our knowledge. Instead of trying to impose an abstract theoretical template (deduction) or 'simply' inferring propositions from facts (induction), we start reasoning at an intermediate level (abduction).

The patterns I observed in the empirical data were regular examples of states taking actions to enhance their legitimacy as actors or to enhance the legitimacy of individual Security Council decisions, which lacks exploration in the literature on Council negotiations. After experimenting with a few different, unsatisfactory, formulations I developed the concept of legitimation practices, which is able to capture the patterns I observed within the empirical material. Legitimation practices also develop a connection between the existing extensive literature on the Security Council and legitimacy with the rapidly growing literature on the Security Council and practice theory.

The term praxiography, meaning practice theory methodology, was coined by Annemarie Mol and shares the thick description aspect of ethnography but with an aim to 'describe and to reconstruct (-graphy), but the interest is not in culture (ethno), but with practice (praxis)' (Bueger and Gadinger 2018: 138).

---

[9] For a detailed description of conducting practice theory-based interviews, see Pouliot (2016: 275–287).

Praxiography describes an approach to the research of practices. Within discussions of praxiography, the most highly regarded method is often participant observation, however, expert interviews and document analysis are also highly regarded approaches and have key advantages when it comes to accessing practices in the security realm (Bueger 2014; Bueger and Gadinger 2018). Participant observation is extremely challenging in the UN Security Council, unsurprisingly, given the security restrictions. To date it has only rarely been conducted by people at the practitioner/academic intersection who had unusual access to private negotiations, such as Michael Barnett (1997, 2002) and Niels Nagelhus Schia (2013, 2017). Given this, I have drawn on elite interviews and primary documents to access the practices used in the negotiations.

This book focuses on the negotiation of resolutions in the Security Council, as they are the preferred format for major decisions which utilize tools such as peacekeeping or sanctions and involve more protracted negotiations than decisions in other forms (such as presidential statements). Resolutions are considered to carry the most weight, and almost all decisions which are intended to be mandatory are issued in this format (Sievers and Daws 2014: 376 and 378).[10] The book therefore focuses on negotiations towards resolutions, although it also includes other forms of decisions where they were steps towards resolutions. After substantive decisions, there are regular follow-up resolutions issued to renew peacekeeping mandates or to renew sanctions committees, however, these tend to be quick and routine in nature, with limited debate. The most substantive and contentious decisions are therefore the first resolutions which authorize the use of different types of coercive tools in relation to a given conflict. As such, this book focuses on the decisions made by the Security Council between 2004 and 2007, as set out in Figure 3, since this is when the Council made its first, highly contentious, decisions. The structure of the book is thematic, around the different 'tools' the UN has authorized in response to the Darfur conflict. The first stage was placing the situation on the Security Council's agenda, followed by negotiations over sanctions, challenging impunity, and then peacekeeping. The thematic chapters are also chronological, as the Security Council considered each tool in sequence, with some overlap.

## Structure of the Book

Part One of the book consists of a background chapter and a theoretical chapter. As background to Security Council negotiations, the first chapter analyses the roles that different actors have within the Council and their capacity to shape

---

[10] Typically, the Security Council refers to Chapter VII of the UN Charter to indicate the binding nature of provisions within a resolution, although this is not always the case (for discussion see Sievers and Daws 2014: 380–393).

| Tool/Chapter | Year of Initial Decisions | Key decisions by resolution |
|---|---|---|
| Agenda-setting | 2004 | • Resolution 1556 established Darfur as an agenda item |
| Sanctions | 2004 to 2006 | • Resolution 1556 imposed an arms embargo on Darfur and threatened further sanctions<br>• Resolution 1564 found Sudan in breach of the earlier resolution but did not authorize further sanctions. It also established an International Commission of Inquiry on Darfur, the findings of which triggered debate on the ICC<br>• Resolution 1591 authorized a sanctions committee to oversee Darfur sanctions<br>• Resolution 1672 issued sanctions against four individuals |
| Referral to the International Criminal Court | 2005 | • Resolution 1593 referred the situation in Darfur to the ICC |
| Peacekeeping | 2006 to 2007 | • Resolution 1679 authorized a Technical Assessment Mission as a precursor to peacekeeping<br>• Resolution 1706 authorized UN peacekeeping in Darfur but could not be implemented due to lack of consent from Khartoum<br>• Resolution 1769 authorized hybrid peacekeeping between the United Nations and the African Union |

**Figure 3**  Key Resolutions on Darfur by the UN Security Council 2004–2007

decision-making. The chapter focuses on the formal and informal power that actors possess via their structural roles within the Council as permanent members, elected members, or as non-voting members of the UN Secretariat. Drawing on Barnett and Duvall's (2005a, 2005b) concept of 'institutional power', I argue that the institutional structures mediate relationships between actors in ways that constrain and enable their capacity to influence negotiations. While the dominance of permanent members is commonly assumed, this chapter explains *how* they dominate, which goes beyond their UN Charter rights of permanency and veto power. Instead, permanent members exercise 'informal veto' on drafts or provisions in drafts, dominate the agenda-setting process, and dominate drafting of decisions.[11] This does not, however, mean that elected members have no capacity to shape negotiations, which is what studies based simply on voting power have argued (Junn 1983; O'Neill 1996). Drafters privilege unanimity, above and beyond the legal requirements, as one internal legitimation practice, which means the votes of elected members are desired for the legitimacy that unanimity bestows. This is especially the case in situations where the elected member is part of the region under discussion. The final way that elected members can shape negotiations is through the diplomatic skill or competency that they bring to the negotiations.[12] While the Secretary-General and Secretariat do not vote on decisions, their legitimated authority still enables them to shape negotiations. The Secretary-General can act as a norm entrepreneur to create an environment where some decisions are valued over others. Members of the UN Secretariat can exert influence in their role as technical advisors to shape peacekeeping mandates (in the case of the Department of Peace Operations) or to shape language used in resolutions (in the case of the Office for the Coordination of Humanitarian Affairs). NGOs, media, and academics, collectively known as the 'third UN', can also shape the normative environment of negotiations as well as applying pressure on specific decisions. This chapter offers a nuanced account of the actors within the Security Council and their different capacities within negotiations.

The second chapter analyses the role of legitimation within Security Council negotiations, representing a shift from the standard approach to analysing the legitimation of the Security Council as an institution or its decisions. Focusing on legitimation practices in negotiations, this chapter introduces the concepts of external legitimation practices and internal legitimation practices. These are 'internal' and 'external' from the vantage point of the Security Council. External legitimation practices are actions taken by states in negotiations to gain the status reward of legitimacy from audiences that matter to them. This chapter addresses two different examples of external legitimation practices. The first is when states prioritize making consistent arguments, even when it leads to an outcome they do not want, such as when the US enabled the referral of Darfur to the ICCC. The second is when states take action via the Security Council to appear as if they

---

[11] Specifically, the United States, the United Kingdom, and France dominate drafting.

[12] For an example of a situation where elected members achieved an outcome through diplomatic skill that permanent members thought they were 'crazy' to even attempt, see (Ralph and Gifkins 2017).

are 'doing something', even when they know the actions cannot be implemented. In both forms of external legitimation practices, states take specific actions within negotiations to appeal to key audiences for the social reward of legitimacy. Internal legitimation practices, on the other hand, are actions taken by drafters to enhance the legitimacy of a decision. Internal legitimation practices include seeking unanimity in negotiations which is above and beyond the legal requirements but is considered the normal goal; seeking support from states which are part of the region under discussion; and using language in decisions which has been 'previously agreed' in other decisions. Internal legitimation practices, while smaller in nature, have nonetheless been instrumental in reaching decisions on Darfur. Internal and external legitimation practices were key in both enabling and shaping decisions on Darfur.

Part Two of the book begins with a chapter on the conflict in Darfur and is followed by four case study chapters: agenda-setting, sanctions, referral to the ICC, and peacekeeping. This thematic progression mirrors the chronological stages of decision-making on Darfur in the Security Council. Chapter 3 sets out the context of the conflict in Darfur, from escalation to widespread violence, and contextualizes debates on both the 'responsibility to protect' and whether the situation in Darfur constituted genocide. It starts with a background on the conflict itself and outlines the early stages—from the initial escalation of the crisis in Darfur in early 2003 until over twelve months later when the Security Council began discussing the situation. During the first six months of the conflict, there was little verified information on the size, scale, and nature of the violence due to the remoteness of the Darfur region. There were few humanitarian agencies on the ground in 2003, and a media blackout orchestrated by the Sudanese government.

While the initial delays were due to lack of information and represent a challenge to the UN's early warning infrastructure, Chapter 4 sets out how the subsequent delays came from Western states resisting Security Council engagement. The UK and US resisted engagement because they were invested in peace talks between the government of Sudan in the north and the southern Sudanese leaders (now South Sudan). There is evidence that the US and the UK had intelligence information detailing the scale of the escalating atrocities in Darfur in 2003 and actively sought to prevent the Security Council's involvement. This calculation shifted when statements from the UN Secretariat in 2004 prompted widespread media interest, which increased domestic interest in Darfur and made it harder for the US and the UK to continue to sideline the situation in Darfur in favour of the North/South peace talks in Sudan. This chapter argues that external legitimation practices enabled initial Security Council engagement with Darfur.

Chapter 5 demonstrates how years of intense negotiations on sanctions were driven by a desire to 'do something', or at least 'to be seen to be doing something', on the part of Western drafters, as a form of external legitimation practices. After two years of intense negotiations, spanning four Security Council resolutions, drafters achieved an arms embargo on the region of Darfur—but not the country of Sudan—and targeted sanctions on only four individuals (Dörfler 2019). Debates on sanctions also show audience sensitivity on the part of China, which repeatedly blocked targeted sanctions in the subsidiary Sanctions Committee but allowed the decision to pass in the Security Council. Beijing said that it would be 'unthinkable' to block the decision in the Security Council when it was supported by all other members, despite China having the legal capacity to do just that (US Embassy Khartoum 2006). As such, China displayed greater sensitivity to the audience perceptions and legitimation implications of vetoing a decision in the Security Council itself, which records public votes, rather than in the subsidiary Sanctions Committee that does not. The desire to 'do something' for Western drafters, and the China's sensitivity against blocking a decision unilaterally, enabled resolutions on sanctions on Darfur to pass.

Chapter 6 analyses the Security Council's decision to refer the situation in Darfur to the ICC. By the time the Security Council began discussing the possibility of this referral in 2005, the US had already spent three years trying to carve out exemptions from prosecution for US personnel via decisions in the Security Council and bilateral non-surrender agreements. As such, the negotiations towards the referral were unique, given the high level of importance they held for a permanent member of the Council. This meant that there was less capacity for legitimation practices to shape these negotiations than there was in the other decisions on Darfur; however, external legitimation practices were still instrumental in the US' decision to abstain and allow the referral to occur. The US had two key goals in relation to the ICC: immunity for US personnel and delegitimizing the ICC as an institution. Widespread support for the ICC, both inside and outside the Security Council, and US advocacy on Darfur, prevented the US from vetoing this resolution, despite a clear statement at the time that it did not want the referral to take place. Internal legitimation practices in the form of repetition of language were also instrumental in facilitating the resolution whereby drafters repeated language from an earlier resolution. Even in this highly contentious case, legitimation practices were instrumental in facilitating a decision and enabling the referral of Darfur to the ICC.

Chapter 7 analyses the drawn-out process of transitioning from African Union peacekeeping in Darfur to UN/AU hybrid peacekeeping, with a failed attempt to implement UN peacekeeping in between. The initial attempt to transfer to UN peacekeeping was driven by the external legitimation practice of 'doing something', even though there were clear statements from Khartoum prior to authorization that they did not consent to UN peacekeeping, which is a necessary precursor for implementation. Two key steps enabled the subsequent authorization of the United Nations–African Union Hybrid Operation in Darfur (UNAMID). First, the UN Secretariat designed the plan from a position of legitimated authority and lobbied both Security Council members and Khartoum to accept it. Second, China shifted position from resisting UN peacekeeping in Darfur to actively lobbying Khartoum to accept UN/AU peacekeeping. This also represented a shift in China's evolving position on peacekeeping and 'consent' whereby it was directly engaged in 'encouraging' this consent. External legitimation practices were central to this shift from China, whereby it sought legitimacy from key international audiences. There were also several internal legitimation practices evident in these negotiations including courting support from African members of the Security Council and a reinvigoration of the drive towards unanimity for resolutions. These internal and external legitimation practices enabled UN/AU peacekeeping to be authorized for Darfur and shaped both the process and outcome of the negotiations.

The concluding chapter is framed by internal and external legitimation practices and shows how each of the key turning points in negotiations on Darfur relate to one or both sets of practices. It shows how decisions on Darfur were facilitated and shaped by internal and external legitimation practices. The chapter shows how this analysis unsettles the often-used image of Western states as advocates on Darfur against Chinese opposition, as the reality was much more nuanced on both sides. It recommends future analysis on internal and external legitimation practices in other case studies or other institutions. It concludes by reflecting on the perennial debates over reform of the Security Council, and how this analysis demonstrates that much of the decision-making process is not an outcome of the formal rules and is therefore open to change without changing the formal rules.

# Bibliography

Adelman, Howard. 2010. 'Refugees, IDPs and the Responsibility to Protect (R2P): The Case of Darfur'. *Global Responsibility to Protect* 2(1–2): 127–148.

Adler, Emanuel and Vincent Pouliot. 2011a. 'International Practices'. *International Theory* 3(1): 1–36.

Adler, Emanuel and Vincent Pouliot. 2011b. 'International Practices: Introduction and Framework'. In *International Practices*, eds. Emanuel Adler and Vincent Pouliot. Cambridge: Cambridge University Press.

Adler-Nissen, Rebecca. 2015. 'Conclusion: Relationalism or Why Diplomats Find International Relations Theory Strange'. In *Diplomacy and the Making of World Politics*, eds. Ole Jacob Sending, Vincent Pouliot, and Ivan B. Neumann. Cambridge: Cambridge University Press.

Adler-Nissen, Rebecca. 2016. 'Towards a Practice Turn in EU Studies: The Everyday of European Integration'. *Journal of Common Market Studies* 54(1): 87–103.

Adler-Nissen, Rebecca and Alena Drieschova. 2019. 'Track-Change Diplomacy: Technology, Affordances, and the Practice of International Negotiations'. *International Studies Quarterly* 63(3): 531–545.

Adler-Nissen, Rebecca and Vincent Pouliot. 2014. 'Power in Practice: Negotiating the International Intervention in Libya'. *European Journal of International Relations* 20(4): 889–911.

Allen, Susan Hannah and Amy T. Yuen. 2013. 'The Politics of Peacekeeping: UN Security Council Oversight Across Peacekeeping Missions'. *International Studies Quarterly* 58(1): 621–632.

Annan, Kofi and with Nadar Mousavizadeh. 2012. *Interventions: A Life in War and Peace*. New York: Penguin.

Badescu, Cristina G. and Linnéa Bergholm. 2009. 'The Responsibility to Protect and the Conflict in Darfur: The Big Let-Down'. *Security Dialogue* 40(3): 287–309.

Barnett, Michael N. 1997. 'The UN Security Council, Indifference, and Genocide in Rwanda'. *Cultural Anthropology* 12(4): 551–578.

Barnett, Michael N. 2002. *Eyewitness to a Genocide: The United Nations and Rwanda* New York: Cornell University Press.

Barnett, Michael N. and Raymond Duvall. 2005a. 'Power in Global Governance'. In *Power in Global Governance*, eds. Michael Barnett and Raymond Duvall. Cambridge: Cambridge University Press.

Barnett, Michael N. and Raymond Duvall. 2005b. 'Power in International Politics'. *International Organization* 59(1): 39–75.

Bashir, Omar S. and Darren J. Lim. 2013. 'Misplaced Blame: Foreign Aid and the Consequences of UN Security Council Membership'. *Journal of Conflict Resolution* 57(3): 509–523.

Beardsley, Kyle and Holger Schmidt. 2012. 'Following the Flag or Following the Charter? Examining the Determinants of UN Involvement in International Crises, 1945–2002'. *International Studies Quarterly* 56(1): 33–49.

Bellamy, Alex J. 2016. 'The Humanisation of Security? Towards an International Protection Regime'. *European Journal of International Security* 1(1): 112–133.

Bellamy, Alex J. and Edward C. Luck. 2018. *The Responsibility to Protect: From Promise to Practice*. Cambridge: Polity Press.

Belloni, Roberto. 2006. 'The Tragedy of Darfur and the Limits of the "Responsibility to Protect"'. *Ethnopolitics* 5(4): 327–346.

Binder, Martin. 2015. 'Paths to Intervention: What Explains the UN's Selective Response to Humanitarian Crises'. *Journal of Peace Research* 52(6): 712–726. online first.

Binder, Martin and Monika Heupel. 2014. 'The Legitimacy of the UN Security Council: Evidence from Recent General Assembly Debates'. *International Studies Quarterly* 59(2): 238–250.

Bode, Ingvild. 2018. 'Reflective practices at the Security Council: Children and Armed Conflict and the Three United Nations'. *European Journal of International Relations* 24(2): 293–318.

Bode, Ingvild and John Karlsrud. 2019. 'Implementation in Practice: The Use of Force to Protect Civilians in United Nations Peacekeeping'. *European Journal of International Relations* 25(2): 458–485.

Bolton, John. 2007. *Surrender is Not an Option: Defending America at the United Nations and Abroad*. New York: Threshold Editions.

Brown, Chris. 2012. 'The "Practice Turn", Phronesis and Classical Realism: Towards a Phronetic International Political Theory?'. *Millennium—Journal of International Studies* 40(3): 439–456.

Buchanan, Allen and Robert O. Keohane. 2011. 'Precommitment Regimes for Intervention: Supplementing the Security Council'. *Ethics and International Affairs* 25(1): 41–63.

Bueger, Christian. 2014. 'Pathways to Practice: Praxiography and International Politics'. *European Political Science Review* 6(3): 383–406.

Bueger, Christian and Frank Gadinger. 2018. *International Practice Theory*, 2nd ed. Basingstoke and New York: Palgrave Macmillan.

Bueno de Mesquita, Bruce and Alastair Smith. 2010. 'The Pernicious Consequences of UN Security Council Membership'. *Journal of Conflict Resolution* 54(5): 667–686.

Caplan, Gerald. 2006. 'From Rwanda to Darfur: Lessons Learned?'. In *Genocide in Darfur: Investigating the Atrocities in the Sudan*, eds. Samuel Totten and Eric Markusen. New York: Routledge.

Chapman, Terrence L. and Dan Reiter. 2004. 'The United Nations Security Council and the Rally "Round the Flag Effect"'. *Journal of Conflict Resolution* 48(6): 886–909.

Charron, Andrea and Clara Portella. 2015. 'The UN, Regional Sanctions and Africa'. *International Affairs* 91(6): 1369–1385.

Claude, Inis L. 1966. 'Collective Legitimization as a Political Function of the United Nations'. *International Organization* 20(3): 367–379.

Clough, Michael. 2005. *Darfur: Whose Responsibility to Protect?*: Human Rights Watch.

Coleman, Katharina P. 2007. *International Organisations and Peace Enforcement: the politics of international legitimacy*. Cambridge: Cambridge University Press.

Cook, Sam. 2016. 'The "woman-in-conflict" at the UN Security Council: a subject of practice'. *International Affairs* 92(2): 353–372.

Cornut, Jérémie. 2017. 'Diplomacy, Agency, and the Logic of Improvisation and Virtuosity in Practice'. *European Journal of International Relations* 24(3): 712–736.

Deplano, Rossana. 2015. *The Strategic Use of International Law by the United Nations Security Council: An Empirical Study*. Springer Briefs in Law. London: Springer.

Dörfler, Thomas. 2019. *Security Council Sanctions Governance: The Power and Limits of Rules*. Routledge Research on the United Nations. Abingdon and New York: Routledge.

Dreher, Axel, Jan-Egbert Sturm and James Raymond Vreeland. 2006. 'Does Membership on the UN Security Council Influence IMF Decisions? Evidence from Panel Data'. *KOF Working Paper 151, ETH Zurich*.

Dreher, Axel, Jan-Egbert Sturm, and James Raymond Vreeland. 2009. 'Development Aid and International Politics: Does Membership on the UN Security Council Influence World Bank Decisions?'. *Journal of Development Economics* 88: 1–18.

Dreher, Axel, Jan-Egbert Sturm, and James Raymond Vreeland. 2015. 'Politics and IMF Conditionality'. *Journal of Conflict Resolution* 59(1): 120–148.

Dunbabin, J.P.D. 2008. 'The Security Council in the Wings: Exploring the Security Council's Non-Involvement in Wars'. In *The United Nations Security Council and War: The Evolution of Thought and Practice since 1945*, eds. Vaughan Lowe, Adam Roberts, Jennifer Welsh, and Dominik Zaum. Oxford: Oxford University Press.

Egeland, Jan. 2008. *A Billion Lives: An Eyewitness Report from the Frontlines of Humanity*. New York: Simon & Schuster.

Eldar, Ofer. 2008. 'Vote-Trading in International Institutions'. *The European Journal of International Law* 19(1): 3–41.

Fasulo, Linda. 2015. *An Insider's Guide to the UN*, 3rd ed. New Haven and London: Yale University Press.

Franck, Thomas M. 1990. *The Power of Legitimacy among Nations*. Oxford: Oxford University Press.

Frederking, Brian and Christopher Patane. 2017. 'Legitimacy and the UN Security Council Agenda'. *PS: Political Science and Politics* 50(2): 347–353.

Friedrichs, Jörg and Fredrich V. Kratochwil. 2009. 'On Acting and Knowing: How Pragmatism Can Advance International Relations Research and Methodology'. *International Organization* 63(4): 701–731.

Frost, Mervyn. 2013. 'Legitimacy and International Organizations: The Changing Ethical Context'. In *Legitimating International Organizations*, ed. Dominik Zaum. Oxford: Oxford University Press.

Frost, Mervyn and Silviya Lechner. 2016. 'Two Conceptions of International Practice: Aristotelian Praxis or Wittgensteinian Language-Games?'. *Review of International Studies* 42(2): 334–350.

Gehring, Thomas and Thomas Dörfler. 2019. 'Constitutive Mechanism of UN Security Council Practices: Precedent Pressure, Ratchet Effect, and Council Action Regarding Intrastate Conflicts'. *Review of International Studies* 45(1): 120–140.

Gifkins, Jess. 2016a. 'Darfur'. In *Oxford Handbook of the Responsibility to Protect*, eds. Alex J. Bellamy and Tim Dunne. Oxford: Oxford University Press.

Gifkins, Jess. 2016b. 'Naming and Framing: Darfur, the Genocide Debate, and the Responsibility to Protect'. In *The United Nations and Genocide*, ed. Deborah Mayerson. Basingstoke: Palgrave Macmillan.

Glennon, Michael J. 2003. 'Why the Security Council Failed'. *Foreign Affairs* 82(3): 16–35.

Gözen Ercan, Pinar. 2022. 'R2P at Twenty: A Shared Responsibility?'. In *The Responsibility to Protect Twenty Years On: Rhetoric and Implementation*, ed. Pinar Gözen Ercan. Cham: Palgrave Macmillan.

Gruber, Lloyd. 2005. 'Power Politics and the Institutionalization of International Relations'. In *Power in Global Governance*, eds. Michael Barnett and Raymond Duvall. Cambridge: Cambridge University Press.

Grünfeld, Fred, Wessel Vermeulen and Jasper Krommendijk. 2014. *Failure to Prevent Gross Human Rights Violations in Darfur: Warnings to and Responses by International Decision Makers (2003–2005)*. Leiden: Brill Nijhoff Publishers.

Guéhenno, Jean-Marie. 2015. *The Fog of Peace: A Memoir of International Peacekeeping in the 21st Century*. Washington D.C.: The Brookings Institution.

Hopf, Ted. 2018. 'Change in International Practices'. *European Journal of International Relations* 24(3): 687–711.

Human Rights Watch. 2004. *Darfur Destroyed: Ethnic Cleansing by Government and Militia Forces in Western Sudan*. Available at http://www.hrw.org/sites/default/files/reports/sudan0504full.pdf.

Hurd, Ian. 1999. 'Legitimacy and Authority in International Politics'. *International Organization* 53(2): 379–408.

Hurd, Ian. 2002. 'Legitimacy, Power, and the Symbolic Life of the UN Security Council'. *Global Governance* 8(1): 35–51.

Hurd, Ian. 2007. *After Anarchy: Legitimacy and Power in the United Nations Security Council*. Princeton: Princeton University Press.

Hurd, Ian. 2008. 'Myths of Membership: The Politics of UN Security Council Reform'. *Global Governance* 14: 199–217.

Hurrell, Andrew. 2005. 'Legitimacy and the Use of Force: Can the Circle Be Squared?'. *Review of International Studies* 31(1): 15–32.

Johnstone, Ian. 2003. 'Security Council Deliberations: The Power of the Better Argument'. *European Journal of International Law* 14(3): 437–480.

Junn, Robert S. 1983. 'Voting in the United Nations Security Council'. *International Interactions: Empirical and Theoretical Research in International Relations* 9(4): 315–352.

Keane, Fergal. 2005. *John Danforth Interview Transcript*. BBC. Available at http://news.bbc.co.uk/2/hi/programmes/panorama/4647211.stm.

Kuus, Merje. 2015. 'Symbolic Power in Diplomatic Practice: Matters of Style in Brussels'. *Cooperation and Conflict* 50(3): 368–384.

Kuziemko, Ilyana and Eric Werker. 2006. 'How Much Is a Seat on the Security Council Worth? Foreign Aid and Bribery at the United Nations'. *Journal of Political Economy* 114(5): 905–930.

Lanz, David. 2020. *The Responsibility to Protect in Darfur: From Forgotten Conflict to Global Cause and Back*. Global Politics and the Responsibility to Protect. Abingdon and New York: Routledge.

Luck, Edward C. 2006. *The UN Security Council: Practice and Promise*. Routledge Global Institutions. Oxon: Routledge.

Malone, David M. 1998. Decision-Making in the UN Security Council: The Case of Haiti, 1990–1997. Oxford: Oxford University Press.

Malone, David M. 2006. *The International Struggle over Iraq: Politics in the UN Security Council 1980–2005*. Oxford: Oxford University Press.

Matsumura, Naoko and Atsushi Tago. 2019. 'Negative Surprise in UN Security Council Authorization: UK and French Vetoes Send Valuable Information for the General Public in Deciding If They Support a US Military Action'. *Journal of Peace Research* 56(3): 395–409.

Mearsheimer, John T. 1994/1995. 'The False Promise of International Institutions'. *International Security* 19(3): 5–49.

Mills, Kurt. 2009. 'Vacillating on Darfur: Responsibility to Protect, to Prosecute, or to Feed?'. *Global Responsibility to Protect* 1(4): 532–559.

Morris, Justin and Nicholas J. Wheeler. 2007. 'The Security Council's Crisis of Legitimacy and the Use of Force'. *International Politics* 44: 214–231.

Neumann, Ivan B. 2002. 'Returning Practice to the Linguistic Turn: The Case of Diplomacy'. *Millennium: Journal of International Studies* 31(3): 627–651.

Nicolini, Davide. 2012. *Practice Theory, Work, and Organisation: An Introduction*. Oxford: Oxford University Press.

Niemann, Holger. 2022. 'Responsibility as Practice: Implications of UN Security Council Responsibilization'. In *The Routledge Handbook on Responsibility in International Relations*, eds. Hannes Hansen-Magnusson and Antje Vetterlein. Abingdon: Routledge.

O'Neill, Barry. 1996. 'Power and Satisfaction in the United Nations Security Council'. *Journal of Conflict Resolution* 40(2): 219–237.

Pouliot, Vincent. 2008. 'The Logic of Practicality: A Theory of Practice of Security Communities'. *International Organization* 62(02): 257–288.

Pouliot, Vincent. 2016. *International Pecking Orders: The Politics and Practices of Multilateral Diplomacy*. Cambridge: Cambridge University Press.

Pouliot, Vincent. 2021. 'The Gray Area of Institutional Change: How the Security Council Transforms Its Practices on the Fly'. *Journal of Global Security Studies* 6(3): 1–18.

Pouliot, Vincent and Jérémie Cornut. 2015. 'Practice Theory and the Study of Diplomacy: A Research Agenda'. *Cooperation and Conflict* 50(3): 297–315.

Prunier, Gerard. 2008. *Darfur: A 21st Century Genocide*, 3rd ed. Crises in World Politics. New York: Cornell University Press.

Ralph, Jason and Jess Gifkins. 2017. 'The Purpose of United Nations Security Council Practice: Contesting Competence Claims in the Normative Context Created by the Responsibility to Protect'. *European Journal of International Relations* 23(3): 630–653.

Rosenthal, Gert. 2017. *Inside the United Nations: Multilateral Diplomacy up Close*. Abingdon and New York: Routledge.

Schia, Niels Nagelhus. 2013. 'Being Part of the Parade—"Going Native" in the United Nations Security Council'. *Political and Legal Anthropology Review* 36(1): 138–156.

Schia, Niels Nagelhus. 2017. 'Horseshoe and Catwalk: Power, Complexity, and Consensus-Making in the United Nations Security Council'. In *Palaces of Hope: The Anthropology of Global Organizations*, eds. Ronald Niezen and Marie Sapignoli. Cambridge: Cambridge University Press.

Schindller, Sebastian and Tobias Wille. 2019. 'How Can We Criticize International Practices?'. *International Studies Quarterly* 63(4): 1014–1024.

Sievers, Loraine and Sam Daws. 2014. *The Procedure of the UN Security Council*, 4th ed. Oxford: Oxford University Press.

Slim, Hugo. 2004. 'Dithering over Darfur? A Preliminary Review of the International Response'. *International Affairs* 80(5): 811–828.

Smith, Courtney B. 2006. *Politics and Process at the United Nations: The Global Dance*. London: Lynn Rienner Publishers.

Spandler, Kilian. 2020. 'UNAMID and the Legitimation of Global-Regional Peacekeeping Cooperation: Partnership and Friction in UN-AU Relations'. *Journal of Intervention and Statebuilding* 14(2): 187–203.

Stedjan, Scott and Colin Thomas-Jensen. 2010. 'The United States'. In *The International Politics of Mass Atrocities: The Case of Darfur*, eds. Paul D. Williams and David R. Black. London and New York: Routledge.

Straus, Scott. 2006. '"Atrocity Statistics" and Other Lessons from Darfur'. In *Genocide in Darfur: Investigating the Atrocities in the Sudan*, eds. Samuel Totten and Eric Markusen. New York: Routledge.

Tallberg, Jonas and Michael Zürn. 2019. 'The Legitimacy and Legitimation of International Organisations: Introduction and Framework'. *The Review of International Organizations* 14: 581–606.

Thompson, Alexander. 2006. 'Coercion Through IOs: The Security Council and the Logic of Information Transmission'. *International Organization* 60(1): 1–34.

Thompson, Alexander. 2010. *Channels of Power: The UN Security Council and US Statecraft in Iraq*. New York: Cornell University Press.

Tzanakopoulos, Antonios. 2014. 'Strengthening Security Council Accountability for Sanctions: The Role of International Responsibility'. *Journal of Conflict & Security Law* 19(3): 409–426.

United Nations. 2005. *5158th Meeting of the United Nations Security Council*. New York, 31 March, S/PV.5158.

United Nations. 2022. UN Security Council Meetings & Outcomes Tables. Dag Hammarskjöld Library. Available at https://research.un.org/en/docs/sc/quick/meetings/2022.

US Embassy Khartoum. 2006. *Sudan/China: UNSC vote on Darfur shows limits of friendship*. Wikileaks. Available at https://wikileaks.org/plusd/cables/06KHARTOUM996_a.html.

Voeten, Erik. 2001. 'Outside Options and the Logic of Security Council Action'. *American Political Science Review* 95(4): 845–858.

Voeten, Erik. 2005. 'The Political Origins of the UN Security Council's Ability to Legitimize the Use of Force'. *International Organization* 59(3): 527–557.

von Billerbeck, Sarah. 2019. '"Mirror, Mirror on the Wall:" Self-Legitimation by International Organizations'. *International Studies Quarterly* 64(1): 207–219.

von Billerbeck, Sarah. 2020. 'No Action Without Talk? UN Peacekeeping, Discourse, and Institutional Self-Legitimation'. *Review of International Studies* 46(4): 477–494.

Welsh, Jennifer. 2016. 'The Responsibility to Prevent: Assessing the Gap between Rhetoric and Reality'. *Cooperation and Conflict* 51(2): 216–232.

Welsh, Jennifer and Dominik Zaum. 2013. 'Legitimation and the UN Security Council'. In *Legitimating International Organizations*, ed. Dominik Zaum. Oxford: Oxford University Press.

Westra, Joel H. 2012. 'Cumulative Legitimation, Prudential Restraint, and the Maintenance of the International Order: A Re-examination of the UN Charter System'. *International Studies Quarterly* 54(2): 513–533.

Wiseman, Geoffrey. 2015. 'Diplomatic Practices at the United Nations'. *Cooperation and Conflict* 50(3): 316–333.

Wood, Michael C. 2016. 'The Interpretation of Security Council Resolutions, Revisited'. *The Max Planck Yearbook on the United Nations Law* 20: 3–35.

Yamashita, Hikaru. 2007. 'Reading "Threats to International Peace and Security" 1946–2005'. *Diplomacy and Statecraft* 18(3): 551–572.

Zaum, Dominik. 2013. 'International Organizations, Legitimacy, and Legitimation'. In *Legitimating International Organizations*, ed. Dominik Zaum. Oxford: Oxford University Press.

# PART ONE

# 1

# Influence in the UN Security Council

*Does the institutional set-up influence United Nations decision making?*
*There is little doubt that the answer to this question is: yes!*

(Kaufmann 1980: 212)

A precursor to understanding how the Security Council handles negotiations for any given conflict or thematic area is to understand how negotiations are conducted in general. This is also useful for highlighting how negotiations for any given topic were convergent or divergent with the usual practices used in negotiations. This chapter analyses each of the key actors in turn, permanent members, elected members, the UN Secretariat, and the broader epistemic community around the United Nations. It shows the forms of influence that each set of actors have in negotiations. The overarching argument made in this chapter—that permanent members dominate negotiations but that elected members and the UN Secretariat, civil society, and media have key avenues for influence—is not new and conforms to common assumptions about the Security Council. As Adler-Nissen and Pouliot have explained, 'While we already knew that the P3 call the shots at the Security Council, our understanding of *how* this is done in actual practice has remained rudimentary' (emphasis original 2014: 909). Where this chapter contributes, however, is via a detailed analysis of how each set of actors—permanent members, elected members, and the UN Secretariat and epistemic community—influence negotiations, and the limits to this influence.

Permanent members dominate agenda-setting and drafting and have the capacity to informally veto provisions, beyond their formal privileges. To explain this, the chapter uses the concept of institutional power, as set out by Barnett and Duvall, which shows how the structure of an institution mediates relationships between participants within the institution (Barnett and Duvall 2005a, 2005b). The influence of permanent members *draws from* the additional privileges they hold under the UN Charter but extends beyond these formal powers. This chapter also demonstrates, however, that internal legitimation practices have the effect of mitigating the dominance of permanent members. Chapter 2 develops the concept of legitimation practices in full, however, for the purposes of this chapter, internal legitimation practices are actions taken by drafters during negotiations to enhance the legitimacy of a decision. Internal legitimation practices mitigate the

*Inside the UN Security Council*. Jess Gifkins, Oxford University Press. © Jess Gifkins (2023).
DOI: 10.1093/oso/9780192869029.003.0002

institutional power held by permanent members, whereby drafters seek unanimity in voting and support from states from the region under discussion. Seeking legitimacy for a decision increases the capacity for influence of elected members. In addition to considerations of legitimacy, elected members can also enhance their influence via acting collectively in groups or via the diplomatic capacities of their individual permanent mission in New York.

While only Security Council members have a vote, this chapter also analyses the forms of influence held by the UN Secretariat, civil society, and the media as non-voting actors. The Secretary-General shapes the normative environment and the conceptual tools that are available to negotiators. Departments of the UN Secretariat also contribute directly to negotiations via briefings or reports and their input carries the weight of legitimated authority. There is growing interest in understanding the different forms of influence that actors have within the UN Security Council, and this chapter contributes to that debate by analysing different forms of influence across voting and non-voting actors (Farrall et al. 2020; Gifkins, Jarvis, and Ralph 2019; Langmore and Farrall 2016; Ralph, Gifkins, and Jarvis 2020). It demonstrates that institutional power is key in explaining how P5 members extend their influence beyond their UN Charter–given privileges, and how internal legitimation practices mitigate institutional power via the legitimacy that E10 support brings to a resolution. Recognizing that the practices of the Security Council are always in flux, the chapter draws from a range of contemporary literature and examples, including but not limited to Darfur, which is the case study in part two of the book.

This chapter fits into the broader book in three key ways. First, it identifies patterns in Security Council negotiations across cases, which both enables perspective for the subsequent case study chapters on Darfur and broadens the analysis of practices beyond the single case study. Second, it identifies internal legitimation practices that enhance the influence of elected members, such as prioritizing unanimity and prioritizing support from states that are located in the region under discussion. These practices incentivize drafters to respond to the concerns of elected members. Third, the chapter identifies a range of ways in which permanent members dominate negotiations; however, internal legitimation practices mitigate the extent of this influence. As such, this chapter serves as background for the later theoretical and empirical material, as well as situating the empirical content in a broader context.

The chapter proceeds in five sections. First, it provides background on the formal structures that guide the Security Council and explains why we need to supplement this with an account of institutional power and legitimation practices. Second, it demonstrates the relationship between the influence of permanent members and institutional power. Third, it shows how internal legitimation practices enhance the influence of elected members as drafters seek legitimacy for decisions in the form of unanimity and regional support. Elected members can also increase their influence by working collectively, and by having strong diplomatic capacities. Fourth, it shows how the UN Secretariat can influence negotiations either by providing advice directly to the Security Council or by shaping the normative environment to encourage specific types of decisions. Finally, it unpacks the influence of the 'third UN'. Where the 'first UN' refers to the United Nations as a forum in which states meet, and the 'second UN' refers to the UN Secretariat, the 'third UN' is made up of the epistemic community surrounding the UN, including NGOs, civil society, academics, and the media. The final section of the chapter analyses the various avenues for influence of the third UN. This chapter demonstrates how different actors influence negotiations and the limits to this influence.

## The Role of Permanent Members

The formal structure of the UN Security Council is deceptively simple. The UN Charter determines that this body has fifteen members[1]: five that hold permanent seats—China, France, Russia,[2] the United Kingdom, and the United States—and ten that hold elected two-year seats. The voting structure magnifies the difference between permanent and short-term positions because it gives each of the permanent members the capacity to block a decision via a negative vote, known as a veto.[3] A vote can also fail if it does not reach the requisite nine affirmative votes. These simple rules form the basis of both membership and voting. In addition, formal rules set out in the Provisional Rules of Procedure govern aspects such as how meetings are called, who can attend meetings, and how the presidency is allocated (United Nations 1982). These formal rules are straightforward; however, they give little insight into how the Security Council operates on a day-to-day basis. The seeming simplicity of these institutional structures belies the rich history of informal practices by which the Security Council makes decisions. Some of these practices are insignificant to decision-making—for example, the Secretary-General always sits to the right of the Council's President at the iconic horseshoe table—which is routine but does not influence decisions (Sievers and Daws 2014: 62). Other informal practices have considerable influence on which decisions are possible—such as which states lead negotiations and how new agenda items are

---

[1] A membership increase from eleven members to fifteen members was authorized via an amendment of the UN Charter in 1963 and came into effect in 1965.

[2] The UN Charter officially states that this seat is held by the Union of Soviet Socialist Republics (USSR), however, in practice, this seat has been occupied by Russia since the break-up of the former USSR.

[3] The term 'veto' does not appear in the UN Charter. Chapter 27(3) calls for the 'concurring' votes of permanent members, however, it was decided early in the Security Council's practice that an abstention was counted as 'concurring' because an abstention was viewed as waiving the right to veto. For discussion, see Bailey (1974).

added. There is a gap, however, between the formal requirements—as set out in the UN Charter and the Provisional Rules of Procedure—and the resulting decisions taken by the Security Council, and this research focuses on the space in between.

Framing P5 roles via institutional power sheds light on the way their power extends beyond their formal privileges. P5 dominance draws from their voting powers but extends beyond voting into other areas. The institutional structure of the Security Council creates greater disparities between permanent and elected members than the formal distinctions between them would suggest (Schia 2013: 139). The only formal difference between permanent and elected members is that permanent members have the capacity to veto decisions and have ongoing membership, while elected members serve two-year terms and cannot individually block decisions. The ongoing nature of P5 membership creates key advantages in the form of institutional memory, enduring relationships, and knowledge of the informal working methods (Langmore and Farrall 2016; Luck 2006; Schia 2017; von Einsiedel, Malone, and Stagno Ugarte 2015). However, the socially sanctioned role that permanent members take up, as an effect of institutional power, is larger than this. For example, since the 1980s the P5 meet separately as a semi-formal practice to coordinate their positions, and they rotate the chair of this group amongst themselves every three months (Sievers and Daws 2014: 126). Specifically, institutional power increases the dominance of permanent members in three key areas: the capacity for 'informal veto', agenda-setting, and penholding, addressed in turn.

First, as an effect of institutional power, permanent members hold an advantage simply by the existence of their veto powers, even without formally exercising them by casting a negative vote. While formal veto votes have been infrequent since the end of the Cold War, the instance of what I term 'informal veto' remains a key potential in negotiations (Schindlmayr 2001: 225). Informal veto occurs when permanent members use the threat of veto to reject provisions of a resolution or an entire resolution during informal negotiations. An example of informal veto comes from the 'straw poll' ballots held when the Security Council selects a new UN Secretary-General. This is not a formal vote but is a 'pre-vote' to indicate voting intentions, wherein permanent members use a different colour to other members, so as to make evident whether a veto would be likely if a formal vote were held for that candidate. In this example, permanent members do not need to issue a negative vote (which would be publicly attached to their name) but can informally veto the proposed individual which then prevents a formal vote on that candidate. Thus, permanent members can use the existence of their veto power to influence the negotiations without the reputational costs of exercising a formal veto. This is an example of how institutional power—mediating relationships between participants—can shape the informal practices of the Council and influence which decisions are possible.

Situations where the threat of a veto blocks a resolution in its entirety are rel-atively rare. A high-profile example of this occurred in the lead-up to the 2003 Iraq War when the US sought a resolution that would formally authorize the use of force. Both Russia and France threatened to veto this so-called 'second resolu-tion', so the US chose not to put this draft forward for a vote (Thompson 2010). The threat of veto more often targets specific provisions in a draft rather than the entirety of a draft. Informal veto has an impact not only when an entire resolu-tion is not put to a vote, but also when key provisions are removed to limit the impact of a resolution which is then passed. As an example of this, in 2019 dur-ing debate towards a resolution on Women, Peace and Security, the US threatened to veto the draft until provisions on sexual and reproductive health for survivors of sexual violence—provisions that had been included in previous resolutions—were removed (Allen and Shepherd 2019). Similarly, the US was able to extract concessions in the draft referral of Sudan to the International Criminal Court (ICC) over crimes committed in Darfur. The US obtained a highly coveted exemp-tion from prosecution for individuals from non-party states to the ICC's founding Rome Statute (excluding Sudan) after threatening to veto the draft (Cryer 2006). In both examples, however, the resolutions passed without key avenues of support or accountability envisaged by drafters. This shows *how* informal veto can affect outcomes: the resolution may pass, but with decreased impact, and can create troubling precedents for subsequent decisions.

Second, permanent members have significant power to determine what is on the Council's agenda (Boulden 2006; United Nations 2004: 79; Wallensteen and Johansson 2004). The capacity to determine the agenda of the Security Council draws from the power to veto decisions held by permanent members. However, the capacity to veto only applies to substantive (non-procedural) questions,[4] while adding items to the agenda is procedural therefore the veto does not apply (Bailey and Daws 1998: 259). In theory, this means that there is no veto right on agenda-setting, but in practice items are only occasionally added to the agenda against the wishes of a permanent member. As a social convention, dominance over agenda-setting is not assumed equally by the P5, but predominantly by the P3, with Russia and China showing little interest in adding items to the agenda (von Einsiedel, Malone, and Stagno Ugarte 2015). For example, the US and the UK resisted efforts to add the situation in Darfur to the agenda of the Security Council through late 2003 and early 2004. They did this because they did not want to jeopardize the concurrent peace negotiations in Sudan between the government of Sudan and the southern region of Sudan (now South Sudan) (Kapila and Lewis 2013; Traub 2010). Being able to influence the agenda means that an actor can use their influ-ence to promote values and practices which are inoffensive to them (Bachrach and Baratz 1962). This also includes 'dogs that don't bark', that is, conflict situations

---

[4] The voting rules come from Article 27 of the United Nations Charter.

that receive limited engagement from the United Nations. For example, conflicts in states which border a P5 member and crises where one or more P5 member is part of the conflict are less likely to have extensive UN involvement than other conflicts (Beardsley and Schmidt 2012). This shows that P5 members can use their positions on the Council to lessen UN involvement in conflicts in which they are directly involved or conflicts in their immediate region. This agenda-setting role is not a formal privilege accorded to permanent members in the UN Charter but is an effect of institutional power wherein the P5, and especially the P3, assume this role for themselves.

Third, institutional power enables permanent members to dominate negotiations via informal penholding practices. Since 2008, agenda items have largely been divided between the P3, with each state assuming the role of 'penholder' on a given issue who then leads the drafting whenever that topic is discussed (Security Council Report 2013: 2). The penholder takes the lead on drafting decisions for a given conflict situation or thematic area and holds political ownership of the topic, meaning that if the penholder does not create a draft it is unlikely that other states will create one (Ralph and Gifkins 2017). By setting the terms of debate, drafters can give other states a choice between X or Y when really they prefer Z (Gruber 2005). The power to interpret a situation is inherently political (Jacobsen and Engell 2018). In particular, penholding has been shown to be one of the primary means of influence within the Security Council (Gifkins, Jarvis and Ralph 2019). As a former German Permanent Representative explained: 'The one who leads, the one who presents the text, who stakes out a position early in the day, is the one who more or less determines the game' (Lynch 2014). In 2022, out of thirty-seven agenda items which had named penholders, P3 members held twenty-seven (73 per cent) (Security Council Report 2022). It is also worth noting that penholding extends beyond drafting and includes, informally, that the penholder organizes open debates, decides whether to hold emergency meetings, and leads visits when the Security Council travels abroad (Security Council Report 2018a: 2). As such, the penholder system gives the P3 considerable power to shape Security Council activity, from debates and events to decisions.

There is no indication of the penholder practice in the UN Charter, and the dominance of the P3 in this role has evolved over time. This is similar to Great Power dominance in the San Francisco negotiations which formalized the UN Charter; smaller states accepted the dominance of Great Powers, however reluctantly, as a cost of getting powerful states to buy in to the institution (Hurd 2007: 95–105). P3 dominance of penholding is thus accepted by other states as an undesirable reality as an effect of the institutional structure. Despite the informal nature of penholding, it has become a key feature of decision-making and the P3 resist attempts to challenge this system (Schia 2017: 61). Permanent members therefore have greater influence over negotiations owing to their dominant role in the drafting of Security Council decisions. As the preceding discussion indicates, there is evidence that institutional power shapes Security Council negotiations towards P5—and especially P3—influence in three main ways: informal veto, agenda-setting, and penholding.

Critically, none of these three phenomena comes directly from the UN Charter or from the Security Council's Provisional Rules of Procedure. Instead, these practices *draw from the existence of* the veto power and permanent membership, but are not explicitly codified in international law and instead come from the way that institutional power mediates relationships between permanent and elected members. These practices have evolved over time and rely on the social acceptance of Security Council members. For example, there is nothing preventing China or Russia from taking a more proactive role in drafting resolutions, yet the P3 dominate penholding. In 2022, China was not penholder for any agenda items, and Russia was only penholder for Central Asia and joint penholder with the US on Golan Heights (Security Council Report 2022). This division of labour is not formally structured by the UN Charter or by the Provisional Rules of Procedure but is accepted, however unwillingly, as 'the way things are done' by elected Security Council members and forms a key part of the informal practices of the Council.

## The Role of Elected Members

The dominance of permanent members could indicate that elected members are of little significance to negotiations. Indeed, one study bluntly concluded that 'a fair approximation as far as voting is concerned is that the Security Council has five members' (O'Neill 1996: 235). Studies by both Junn (1983) and O'Neill (1996) argue that each individual elected member of the Council has almost no power, if you conceptualize 'power' as the ability of an individual state to determine the outcome of a particular vote. They both base their arguments on statistical models of formal voting rules, and even though the power that permanent members wield extends beyond these formal rules, elected members can still have influence in negotiations. Critically, shared conceptions of legitimacy mitigate the institutional power that P5 members enjoy, which gives elected members a more significant role than their voting power alone suggests. Recognition of the centrality of legitimacy to the functioning of the Security Council is not new; however, the way legitimacy shapes negotiations needs further attention (Claude 1966; Hurd 2007; Thompson 2010). Internal legitimation practices—used by drafters—enhance the influence of elected members. The concept of legitimation practices is developed in full in Chapter 2; however, for the purposes of this chapter, there are two forms of internal legitimation practices which increase the agency of elected members and their capacity to influence negotiations. First, drafters prioritize unanimity in decision making, so elected members can leverage this to enhance their influence (Dunne and Gifkins 2011: 523). Second, if an elected member is from the region under discussion, the 'legitimacy value' of its support increases, which can also be leveraged for influence. In each of these scenarios, the symbolic power of support from elected members increases, which affords greater scope for leverage by elected members (Hurd 2002). I outline these two internal legitimation practices below, and their relationship with influence.

First, unanimous decision-making is highly valued within the Security Council, so any threat to unanimous voting can increase the potential influence of a dissenting state.[5] Legally, the requirement to pass a resolution is only nine affirmative votes, if there are no negative votes from permanent members. Yet the Security Council displays a strong preference for unanimous voting. The vast majority of resolutions put to a vote in the Security Council since the end of the Cold War have not only passed, but pass unanimously (Dunne and Gifkins 2011: 523). Between 2000 and 2021, there were 1386 draft resolutions voted on in the Security Council, of which 87 per cent not only passed but also had the support of all fifteen members.[6] Unanimous voting is recognized as a mark of legitimacy within the Security Council and is seen to strengthen a decision and the likelihood that it will be implemented. For example, the former UN Department of Peacekeeping Operations' (DPKO) Capstone Report lists unity among the key attributes which influence the success of a peacekeeping operation: 'anything other than unanimous Security Council backing can be a serious handicap' (DPKO 2008: 50). This drive towards unanimity creates more space for E10 influence. For example, after Sudan rejected a peacekeeping mission in 2006, Security Council members realized that it was particularly important that the next resolution on peacekeeping in Darfur was unanimous. The meeting record for the unanimous adoption of subsequent resolution 1769 includes statements from both Qatar and Indonesia, showing that drafters accommodated their concerns in the draft so that they were able to vote for the resolution (United Nations 2007). While Junn and O'Neill are right that individual elected members have limited power to determine the outcome of a vote, they can generate more influence in exchange for an affirmative vote due to the legitimation practices of drafters and the drive towards unanimity.

Second, the legitimacy value of support from elected members is not equal between elected members; drafters especially prize support from states that are within the region under discussion. This support could be in the form of affirmative votes, co-sponsorship, or any other leadership role. For example, a draft resolution on Syria in 2012 was initially led by the P3 before it was presented to the Council by Morocco, as an Arab state on the Council (Gifkins 2012: 385). The think tank Security Council Report explained at the time of this draft: 'Arab support is seen as crucial in getting a unanimous Council position on the Syria

---

[5] There can also be some pressure to vote affirmatively which is an area that warrants further research (see for example Schia 2013; Adler-Nissen and Pouliot 2014: 900).

[6] While the average remains high, between 2018 and 2020 the average dropped to 74 per cent, although it then increased to 81 per cent in 2021. This data was compiled by the author (UN 2022).

crisis' (Security Council Report 2012). As such, Arab leadership on the draft was an internal legitimation practice that helped to secure wider support. Similarly, a leaked US cable discusses a strategy wherein support from Ghana was sought for a draft resolution on Darfur as a means to subsequently secure support from China, Qatar, and South Africa (US Permanent Mission to the UN 2007). These examples show the value of regional support as a legitimation practice used by drafters to enhance wider support. The elected member in question can then translate their support into influence. One diplomat who was involved in the negotiations towards resolution 1973 on Libya, when Lebanon acted as a conduit between the Arab League and the Security Council, explained the approach thus: 'this is our patch, and we are telling you that you should do X, Y and Z' (Adler-Nissen and Pouliot 2014: 901). It was at the suggestion of Lebanon that the text 'no foreign occupation' was added into resolution 1973 as a political compromise that helped to facilitate support for the resolution (Adler-Nissen and Pouliot 2014: 901). Regions are increasingly seen as playing a 'gatekeeper' role by Bellamy and Williams, and while they specifically discussed support from regional groups, support from regionally relevant states is also prized in Security Council negotiations (2011).[7] Support from states in the region is valued for the legitimacy it brings to a draft.

Internal legitimation practices can increase the capacity for influence that elected members can have in negotiations. If drafters want unanimity and support from the region, there is an incentive to accommodate the priorities of elected members. These are not the only means of enhancing the influence of elected members, however. Elected members can also increase their capacity for influence via strategic means, either by collaborating with fellow elected members to act as a bloc, or by having strong diplomatic capacities in terms of their competencies in Security Council practices and the size and effectiveness of their diplomatic corp. These two strategies can give elected members greater input into drafting, or in some cases create opportunities for elected members to lead negotiations themselves. I discuss each of these strategies below.

First, there is currently growing recognition from elected members that they can magnify their influence by acting collectively. This development is a response to the frustrations of elected members at the dominance of permanent members in drafting. When a group of elected members share a common position on an issue, they can negotiate as a bloc, which enhances their voting power. As Gowan explained, 'A nascent "E10 culture" has emerged in recent years, with small and medium-sized countries working together across regional divides to make their presence felt' (2019). Elected Security Council members outlined their

---

[7] In addition to regional leadership, there are also instances where informal groups form to address a specific conflict either inside or alongside the UNSC, such as the 'P5 + 1' group including Germany, which works on negotiations on Iran (see Prantl 2005).

intentions to work more collectively in a series of documents in 2018 (Sievers and Daws 2018). E10 members often have divergent foreign policies, but they have drawn strength in working collectively on topics where they can find common ground. For example, on Yemen, a group of five elected members worked together to submit joint proposals, and even threatened to 'take the pen' from the UK, which is the penholder on Yemen (Martin 2018; Security Council Report 2019). Similarly, a different group of five elected members worked together to draft a resolution condemning attacks against medical facilities and personnel (Martin 2018). Another prominent example is a revolving group of elected members that have taken over penholding for the humanitarian track of the Syrian conflict. The group was initially Australia, Luxemburg, and Jordan and has evolved as elected members have rotated on and off the Security Council (Langmore and Farrall 2016; Ralph and Gifkins 2017). Another group of elected members seized the pen for a resolution on Israeli settlements and by doing so were able to avoid a US veto (Loiselle 2020). While these small groups of elected members are often not numerically large enough to block a decision, elected members have recognized the power of acting in collectively—where possible—to magnify their influence.

Second, elected members can increase their influence via their diplomatic capacity. The duration of tenure for permanent members means that they benefit from institutional memory and detailed knowledge of procedures and precedents (Farrall et al. 2020; Security Council Report 2018b). Even experienced UN diplomats describe a steep learning curve when taking up an elected position in the Security Council, while the P5 are 'old hands, comfortable in their skins' (Rosenthal 2017: 94–95). Elected members can mitigate their institutional disadvantage via diplomatic capacity, meaning both competence in Security Council practices and the size and reach of their diplomatic infrastructure.

Diplomatic capacities have been shown to be key to a states' capacity for influence in the Security Council, both in general and for elected members specifically (Farrall et al. 2020; Gifkins, Jarvis and Ralph 2019). States taking up a seat in the Council can ramp up their skills and knowledge, such as Sweden's strategy of building a database of agenda items and positions of states, and Norway's strategy of drawing on knowledge via partnerships with universities and think tanks (Thorhallsson 2012). Competence in diplomatic practices can enhance the ways in which members are able to influence decision-making. For example, Australia used diplomatic skill and constructive relationships to secure a resolution calling for an investigation following the downing of Malaysian Airlines flight MH17, despite Russian resistance to this proposal (Langmore and Farrall 2016). Elected members can benefit from engaging in 'niche diplomacy' during their tenure, and using their presidency to facilitate debates on favoured topics can be a useful strategy (Lupel and Mälksoo 2019). An area where elected members dominate is chairing Sanctions Committees, which are a subsidiary body of the Security

Council, giving them influence over how sanctions are applied (Eckert 2016).[8] Elected members can also have more targeted impact in thematic agendas, such as Japan's leadership role on working methods (Pouliot 2021). Through diplomatic skill and activism, some elected members can and do 'punch above their weight'. Diplomatic capacity can increase the influence that a state has in decision-making while serving on the Council.

By way of summary then, the influence of E10 states in Security Council negotiations increases with the internal legitimation practices of drafters, or the diplomatic capacities of the elected member state. Drafters prioritize unanimity as a standard practice for the legitimacy it brings to a decision. An abstention or negative vote of an elected member rarely has any legal impact on the outcome of the vote, but drafters nonetheless consider this undesirable. The support of elected members from the region under discussion carries additional weight in terms of the legitimacy it brings to a decision. States can alternatively increase their influence in negotiations via their strong diplomatic capacities or by acting as a bloc. While elected members are highly unlikely to be in a position where they can individually determine the outcome of a decision, they do have more influence in decision-making than is suggested by statistical models based purely on voting.

The scenarios and attributes that can increase E10 agency should not be over-stated, however. Much of the decision-making, particularly on contentious issues, occurs among permanent members. The timing of negotiations illustrates this, whereby the P3 agrees drafts then negotiate them with Russia and China. Only then, once an often delicate balance has been reached, are drafts shared with elected members, with limited time before a vote and discouragement from drafters for substantive change (von Einsiedel, Malone and Stagno Ugarte 2015). As described by Nadin, by the time drafts are circulated to the E10, they have 'a de facto disclaimer: do not call for amendments that might upset the consensus achieved among permanent members' (2016: 108). E10 members describe frustration at a role that often feels like providing a 'rubber stamp' to decisions that have already been made by permanent members (Ralph and Gifkins 2017: 643). While internal legitimation practices and diplomatic capacities can increase the influence of elected members, this remains limited in a system that is weighted towards permanent members, which dominate drafting and can individually block decisions.

## The Role of the UN Secretary-General and Secretariat

The other set of actors with direct input into Security Council negotiations are the UN Secretary-General and the Secretariat more broadly. This group of actors

---

[8] This chapter does not focus on the Security Council's subsidiary bodies. For further discussion on the decision-making process in Sanctions Committees see Gehring and Dörfler (2013).

are sometimes known as the 'second UN' where the 'first UN' refers to the inter-governmental fora (Weiss et al. 2007). The Secretariat does not have a vote in Security Council negotiations but has the capacity to influence decision-making, nonetheless. The Secretariat has legitimated authority which enhances its capacity for influence (Cronin and Hurd 2008). The Secretary-General can influence which decisions are possible by shaping the normative environment. By doing so the Secretariat can create a linguistic and political repertoire, which encourages normative change and the development of new tools. The other way that the second UN can shape decisions is by providing information and advice that informs decisions. International organizations can act as seemingly neutral information providers with legitimated authority (Abbott and Snidal 1998). In this way information provided by Secretariat departments is often held in high esteem, even though institutions can have their own political interests that can drive their advice (Barnett and Finnemore 2004). Former head of the Department of Peace Operations (DPO)[9], Jean-Marie Guéhenno, has said of the Secretariat 'It actually has much more power than is generally assumed, and except for the few crises where a major power has a strategic interest, the Secretariat can often be in the driver's seat, like a sixth permanent member of the council' (Guéhenno 2015: 312). Despite not having a vote in Security Council decisions, the UN Secretary-General and Secretariat can shape decisions directly by providing information and recommendations or indirectly by promoting a normative environment that enables particular types of decisions, both of which are discussed in turn below.

First, the UN Secretary-General can provide normative guidance for the Security Council both via norm entrepreneurship and as a moral leader. The office of the Secretary-General has been described as 'responsibility without power', however, it does have capacity for influence (Urquhart 2007: 31). The Secretary-General's political role comes from the existence of Article 99 of the UN Charter which enables her or him to bring matters to the attention of the Security Council (United Nations 1945). This provision is rarely formally invoked, but it increases the political capacity of the Secretary-General's role (Kille 2005). UN Secretaries-General can increase their influence by acting as a 'norm entrepreneur', developing and championing new concepts. The office of the UN Secretary-General has a unique capacity for norm entrepreneurship (Adebajo 2007; Johnstone 2007; Rushton 2008). Norm entrepreneurs frame issues in innovative ways and shape the agenda in new directions (Finnemore and Sikkink 1998: 897). Prominent examples of past Secretaries-General acting as norm entrepreneurs are: Dag Hammarskjold and his advocacy of peacekeeping, Boutros Boutros-Ghali's advocacy on democratization, and Kofi Annan and his advocacy on the responsibility to protect (Johnstone 2007). By contributing to the normative environment in the

---

[9] The Department of Peace Operations was called the Department of Peacekeeping Operations at the time.

Security Council, the Secretary-General can help shape the conceptual tools that are available to states when making decisions on matters of international peace and security. Secretaries-General can also have normative influence by providing moral leadership. Although they cannot tell states what to do, 'it is the job of the Secretary-General to make it harder for an international community to make manifestly wrong decisions, or to take no decision at all' (Chesterman and Franck 2007: 240). Secretaries-General develop the conceptual tools that decision makers can use, and can provide moral leadership to encourage certain decisions from states.

Second, the Secretary-General and departments of the UN Secretariat also shape individual decisions by providing expert recommendations that inform deliberations. The Secretary-General and Secretariat are perceived as technical experts with impartiality which gives their recommendations unique gravitas (Barnett and Finnemore 2004). For example, the DPO has a unique role in planning peacekeeping operations and advising the Security Council on resources, mandate, and strategy. The Security Council uses the DPO as impartial experts: 'It is the [DPO] planners themselves who frame the options for UN engagement that are presented to the Security Council by the Secretary-General, right down to the mandate language. Although the Council can ignore or modify the Secretary-General's recommendations when authorizing peacekeeping missions, it typically follows his [sic] guidance closely' (Holt, Taylor and Kelly 2009: 119). By providing recommendations for the Security Council, the DPO can shape the Security Council's response and its interpretation of the options available. Other Secretariat departments also guide Security Council decision-making. The Office for the Coordination of Humanitarian Affairs (OCHA) produces a regular Aide Memoire for the Consideration of Issues Pertaining to the Protection of Civilians in Armed Conflict (see for example 2014a, 2016). While this document is based on the past practices of the Security Council it can also be seen as an advocacy tool, as OCHA has identified the lack of consistency in Council language as a problem and it advocates greater consistency (OCHA 2014b). By drawing together past language from Security Council resolutions into an accessible form OCHA can encourage the Council towards more consistent use of language; thus while acting impartially it can also promote its goal of greater consistency. The technical expertise of the UN Secretariat gives it some ability to shape both the process and the outcome of negotiations towards its preferred outcomes.

The above roles of providing expert advice and shaping the normative environment may give an overly rosy image of the Secretariat, however, and it is worth noting that entities within the Secretariat also have their own interests, which can lead to dysfunction. The position taken by the UN Secretariat during the Rwandan genocide is an oft-repeated example of this, with good reason. There was a UN peacekeeping mission in Rwanda at the time the genocide began, to monitor a previous peace agreement. When violence escalated the violence was interpreted through the frame of a 'return to civil war', which meant that there was 'no peace to keep', and the peacekeeping operation was reduced (Barnett and Finnemore 2004). Earlier failures in Bosnia and Somalia were seen as a threat to the organization, which meant that the Secretariat was increasingly risk-averse (Barnett and Finnemore 2004). Michael Barnett makes a compelling argument that the Secretariat prioritized protecting the UN over protecting Rwandans, in part by withholding information from the Security Council (1997, 2002). He concludes that, 'My view is that the UN bears some moral responsibility for the genocide' (2002: 20). In this example, the interests of the Secretariat as an actor took precedence.

Beyond Rwanda, another example is the UNHCR focus on 'declaring' a country safe for return, whereby 'voluntary repatriation' shifted from consent by refugees to an assessment criteria by UNHCR enabling it to send Rohingya refugees back to Myanmar (Barnett and Finnemore 2004: 100–110). The UN Secretariat, like other bureaucracies, are known 'for acting in ways that are at odds with their stated mission, and for refusing requests of and turning their backs on those to whom they are officially responsible' (Barnett and Finnemore 1999: 715). Autesserre has also shown how practitioners in what she terms 'Peaceland', which could include the UN Secretariat, have their own logics, knowledges, and practices, and that these can be ineffective and counterproductive (2014). As such, while the Secretariat can influence Security Council decisions via norm entrepreneurship or providing advice, this influence is not necessarily normatively advantageous and indeed can be dysfunctional.

## The 'Third UN'

The preceding sections have analysed the 'first UN', which is the 'forum' image of the United Nations in the form of permanent and elected members of the Security Council,[10] and the 'second UN' in the form of the UN Secretariat. The 'third UN' is a collection of actors and individuals who are not representing their country (like the first UN), nor are they employed by the UN Secretariat (like the second UN). The third UN includes NGOs, civil society, academics, media representatives, and consultants, who are closely associated with the UN but are not formally part of it (Weiss, Carayannis, and Jolly 2009; Weiss et al. 2017). NGOs, and civil society more broadly, have long been recognized as having a key role in shaping the normative environment in which UN decisions are made (Risse and Sikkink 1999). NGOs also play key roles in gathering and disseminating information to and about the UN, of which Security Council Report is an important actor closely

---

[10] The UN General Assembly, Human Rights Council, Economic and Social Council, and other UN forums also make up the 'first UN' where states meet.

monitoring current Council negotiations as well as the politics of decisions that have been made (von Einsiedel, Malone and Stagno Ugarte 2015). There are limits to the influence of NGOs, however, and what can be said, along with how it can be said, are constrained by informal norms around acceptable language and framing of issues (Gibbings 2011). Despite these limitations, NGOs, and broader civil society movements, play an important role in providing information to the Security Council via Arria formula meetings,[11] framing situations, generating salience, and lobbying the Security Council and its members to do more and act faster.

Academics and the media can also shape debates inside the United Nations, by providing advice, shaping debates, and shaping the normative environment. Academics working alongside the United Nations 'have had an impact in fostering ideas and policies, including human development, climate change, global compact, sovereignty as responsibility, and human security' (Kittikhoun and Weiss 2011: 18). The UN also has a specific body to facilitate the exchange of ideas between academics and UN staff in the form of the United Nations Academic Impact organization, which is a network of higher education institutions. Similarly, the media plays a significant role in framing public understandings of complex conflicts. Comparative media studies comparing media coverage of the Syrian conflict by Russian and Western outlets, and comparing coverage of the Darfur conflict by Chinese and Western outlets, found significant differences in the amount of coverage and timing of the coverage, the framing of the conflict, the emotional intensity of the reporting, and the nature of the violence (Brown 2014; Mody 2010). Other studies have highlighted the role of visual politics in news photography and how images serve to emphasize certain framings and provoke particular emotional resonance (Campbell 2007). As such, there is scope for academics and media representatives to advise, lobby, inform, and shape the understanding of conflicts, both inside and outside the United Nations.

## Conclusion

The formal rules that govern the UN Security Council—the UN Charter and the Provisional Rules of Procedure—offer little insight into the day-to-day process of negotiating decisions. Analysing the informal, and often unwritten practices, is necessary to understand decision-making and the types of influence that different actors have. This chapter demonstrates the utility of stripping back analysis of the Security Council to the roles and capacities of actors within negotiations. By considering the roles of actors within negotiations—permanent members, elected

---

[11] 'Arria formula' meetings, named after the former Permanent Representative of Venezuela Diego Arria, are informal gatherings of Security Council members (rather than formal meetings) which enable the Council to be briefed by people from outside the UN system, primarily NGOs (Sievers and Daws 2014: 74–76).

members, the UN Secretariat, and the 'third UN'—the different forms of influence that they have come to the fore. While there is often an assumption that permanent members dominate decision-making, it has been lamented that there is still limited understanding of how this dominance occurs (Adler-Nissen and Pouliot 2014). The framing of institutional power helps to explain how permanent members extend their power far beyond their formal privileges—via the way formal institutional structures mediate relationships between participants. As such, permanent members gain greater informal influence in drafting, agenda-setting, and their capacity for informal veto, despite these features being informal practices.

Legitimation practices mitigate the dominance of permanent members, however. While individual elected members have almost no power to determine the outcome of a negotiation, they are able to have influence due to the drive towards unanimity, or as states from the region under discussion. Drafters seek the legitimacy that comes from E10 support, and elected states can leverage that for influence. Individual E10 members also have greater influence when they work collectively or if they have strong diplomatic capacities. While only fifteen states can vote in the Security Council, the UN Secretariat can also have influence by shaping the normative environment or providing recommendations from a position of legitimated authority. NGOs, civil society, academics, and the media can also influence Security Council negotiations by providing information, lobbying, framing issues, and promoting broader salience of issues. By stripping back analysis of the Security Council to the roles of actors inside negotiations, this chapter provides the necessary background on the Security Council as an institution, the impact of institutional power, and begins to introduce the role of internal legitimation practices in mitigating institutional power.

The argument on how dominance and influence occur also has implications for the perennial debates over Security Council reform. As has been well established, the debates over formal reform—that is, change to the membership or voting rules set out in the UN Charter—are deadlocked (Nadin 2016; Niemetz 2015).[12] As such it has been argued that the most promising avenues for reform are informal changes to practices or working methods (Niemetz 2015; Wenaweser 2016). The Security Council as an institution is continually adapting, which means that the above informal practices are not set in stone but can change over time (Luck 2016; MacKenzie 2015; Yamashita 2007). The gains in recent years made by elected members working collaboratively in small groups—for example, on the Syrian humanitarian track—shows that there is potential for elected members to take the initiative and make an impact where they have strong diplomatic skills and a narrowly targeted goal. Since 2018, elected members have sought to codify

---

[12] Formal amendment of the UN Charter requires ratification from two-thirds of the UN General Assembly and all five permanent members of the UN Security Council. While there is widespread agreement on the need for reform, there is no single model of reform which currently has enough support.

their intentions to collaborate, and while this is not a panacea, it does show that there are avenues for greater capacity for influence by elected members. There has also been growing frustration with the dominance of penholding by P3 members. A welcome step from the UK in 2019 was to co-penhold with Germany on Libyan sanctions, where Germany is the Chair of the Libyan Sanctions Committee (Security Council Report 2019). Elected members gain specific expertise on sanctions during their tenure, as elected members chair most Sanctions Committees, and expanding co-penholding on sanctions or other areas would enable elected members to have a more active role in drafting. These avenues for increased influence of elected members show that, while formal reform debates are deadlocked, there is potential for informal change that could facilitate broader participation in negotiations, away from the P3 stranglehold. As such, this chapter demonstrates how the formal rules make up only a small component of the distribution of power and influence within the Security Council. The UN Charter and Provisional Rules of Procedure give structure to the Security Council but reveal little about how decisions are made; a process which has evolved over decades into an environment rich with informal practices.[13]

# Bibliography

Abbott, Kenneth W. and Duncan Snidal. 1998. 'Why States Act through Formal International Organizations'. *The Journal of Conflict Resolution* 42(1): 3–32.

Adebajo, Adekeye. 2007. 'Pope, Pharaoh, or Prophet? The Secretary-General after the Cold War'. In *Secretary or General? The UN Secretary-General in World Politics*, ed. Simon Chesterman. Cambridge: Cambridge University Press.

Adler-Nissen, Rebecca and Vincent Pouliot. 2014. 'Power in practice: Negotiating the international intervention in Libya'. *European Journal of International Relations* 20(4): 889–911.

Allen, Louise and Laura Shepherd. 2019. *In pursuing a new resolution on sexual violence Security Council significantly undermines women's reproductive rights*. London School of Economics. Available at https://blogs.lse.ac.uk/wps/2019/04/25/in-pursuing-a-new-resolution-on-sexual-violence-security-council-significantly-undermines-womens-reproductive-rights/.

Autesserre, Severine. 2014. *Peaceland: Conflict Resolution and the Everyday Politics of International Intervention*. New York: Cambridge University Press.

Bachrach, Peter and Morton S. Baratz. 1962. 'Two Faces of Power'. *American Political Science Review* 56(4): 947–952.

Bailey, Sydney D. 1974. 'New Light on Abstentions in the UN Security Council'. *International Affairs* 50(4): 554–573.

Bailey, Sydney D. and Sam Daws. 1998. *The Procedure of the UN Security Council*, 3rd ed. Oxford: Oxford University Press.

---

[13] An earlier version of this background chapter was published in *Global Governance* as Jess Gifkins (2021) 'Beyond the Veto: Roles in UN Security Council Decision-Making'. *Global Governance* 27(1): 1–24.

Barnett, Michael N. 1997. 'The UN Security Council, Indifference, and Genocide in Rwanda'. *Cultural Anthropology* 12(4): 551–578.

Barnett, Michael N. 2002. *Eyewitness to a Genocide: The United Nations and Rwanda* New York: Cornell University Press.

Barnett, Michael N. and Raymond Duvall. 2005a. 'Power in Global Governance'. In *Power in Global Governance*, eds. Michael Barnett and Raymond Duvall. Cambridge: Cambridge University Press.

Barnett, Michael N. and Raymond Duvall. 2005b. 'Power in International Politics'. *International Organization* 59(1): 39–75.

Barnett, Michael N. and Martha Finnemore. 1999. 'The Politics, Power, and Pathologies of International Organizations'. *International Organization* 53(4, Autumn): 699–732.

Barnett, Michael N. and Martha Finnemore. 2004. *Rules for the World: International Organizations in Global Politics*. Ithaca and London: Cornell University Press.

Beardsley, Kyle and Holger Schmidt. 2012. 'Following the Flag or Following the Charter? Examining the Determinants of UN Involvement in International Crises, 1945–2002'. *International Studies Quarterly* 56(1): 33–49.

Bellamy, Alex J. and Paul D. Williams. 2011. 'The New Politics of Protection? Cote d'Ivoire, Libya and the Responsibility to Protect'. *International Affairs* 87(4): 845–870.

Boulden, Jane. 2006. 'Double Standards, Distance and Disengagement: Collective Legitimization in the Post-Cold War Security Council'. *Security Dialogue* 37(3): 409–423.

Brown, James D.J. 2014. "Better One Paper Tiger than Ten Thousand Rapid Rats': Russian Media Coverage of the Syrian Conflict'. *International Politics* 50(1): 45–66.

Campbell, David. 2007. 'Geopolitics and Visuality: Sighting the Darfur Conflict'. *Political Geography* 26: 357–382.

Chesterman, Simon and Thomas M. Franck. 2007. 'Resolving the Contradictions of the Office'. In *Secretary or General? The UN Secretary-General in World Politics*, ed. Simon Chesterman. Cambridge: Cambridge University Press, pp. 232–240.

Claude, Inis L. 1966. 'Collective Legitimization as a Political Function of the United Nations'. *International Organization* 20(3): 367–379.

Cronin, Bruce and Ian Hurd. 2008. 'Introduction'. In *The UN Security Council and the Politics of International Authority*, eds. Bruce Cronin and Ian Hurd. Oxon: Routledge.

Cryer, Robert. 2006. 'Sudan, Resolution 1593, and International Criminal Justice'. *Leiden Journal of International Law* 19(1): 195–222.

DPKO. 2008. *United Nations Peacekeeping Operations: Principles and Guidelines*. United Nations. Available at https://peacekeeping.un.org/sites/default/files/capstone_eng_0.pdf.

Dunne, Tim and Jess Gifkins. 2011. 'Libya and the State of Intervention'. *Australian Journal of International Affairs* 65(5): 515–529.

Eckert, Sue. 2016. 'The Role of Sanctions'. In *The UN Security Council in the 21st Century*, eds. Sebastian Von Einsiedel, David M. Malone, and Bruno Stagno Ugarte. Boulder and London: Lynne Rienner.

Farrall, Jeremy, Marie-Eve Loiselle, Christopher Michaelsen, Jochen Prantl, and Jeni Whalen. 2020. 'Elected Member Influence in the United Nations Security Council'. *Leiden Journal of International Law* 33(1): 101–115.

Finnemore, Martha and Kathryn Sikkink. 1998. 'International Norm Dynamics and Political Change'. *International Organization* 52(4): 887–917.

Gehring, Thomas and Thomas Dörfler. 2013. 'Division of Labor and Rule-based Decisionmaking within the UN Security Council: The Al-Qaeda/Taliban Sanctions Regime'. *Global Governance* 19(4): 567–587.

Gibbings, Sheri Lynn. 2011. 'No Angry Women at the United Nations: Political Dreams and the Cultural Politics of United Nations Security Council Resolution 1325'. *International Feminist Journal of Politics* 13(4): 522–538.

Gifkins, Jess. 2012. 'The UN Security Council Divided: Syria in Crisis'. *Global Responsibility to Protect* 4(3): 377–393.

Gifkins, Jess. 2021. 'Beyond the Veto: Roles in UN Security Council Decision-making'. *Global Governance* 27(1): 1–24.

Gifkins, Jess, Samuel Jarvis, and Jason Ralph. 2019. 'Brexit and the UN Security Council: Declining British Influence?' *International Affairs* 95(6): 1349–1368.

Gowan, Richard. 2019. *Council of Despair? The Fragmentation of UN Diplomacy.* International Crisis Group. Available at https://www.crisisgroup.org/global/b001-council-despair-fragmentation-un-diplomacy.

Gruber, Lloyd. 2005. 'Power Politics and the Institutionalization of International Relations'. In *Power in Global Governance*, eds. Michael Barnett and Raymond Duvall. Cambridge: Cambridge University Press.

Guéhenno, Jean-Marie. 2015. *The Fog of Peace: A Memoir of International Peacekeeping in the 21st Century.* Washington D.C.: The Brookings Institution.

Holt, Victoria, Glyn Taylor, and Max Kelly. 2009. *Protecting Civilians in the Context of UN Peacekeeping Operations: Successes, Setbacks and Remaining Challenges.* New York: United Nations.

Hurd, Ian. 2002. 'Legitimacy, Power, and the Symbolic Life of the UN Security Council'. *Global Governance* 8(1): 35–51.

Hurd, Ian. 2007. *After Anarchy: Legitimacy and Power in the United Nations Security Council.* Princeton: Princeton University Press.

Jacobsen, Katja Lindskov and Troels Gauslå Engell. 2018. 'Conflict Prevention as Pragmatic Response to a Twofold Crisis: Liberal Interventionism and Burundi'. *International Affairs* 94(2): 363–380.

Johnstone, Ian. 2007. 'The Secretary-General as Norm Entrepreneur'. In *Secretary or General? The UN Secretary-General in World Politics*, ed. Simon Chesterman. Cambridge: Cambridge University Press.

Junn, Robert S. 1983. 'Voting in the United Nations Security Council'. *International Interactions: Empirical and Theoretical Research in International Relations* 9(4): 315–352.

Kapila, Mukesh and Damien Lewis. 2013. *Against a Tide of Evil: How one man became the whistleblower to the first mass murder of the twenty-first century.* Edinburgh and London: Mainstream Publishing.

Kaufmann, Johan. 1980. *United Nations Decision Making*, 3rd ed. Alphen aan den Rijn: Sijthoff & Noordhoff.

Kille, Kent J. 2005. *From Manager to Visionary: The Secretary-General of the United Nations.* New York: Palgrave Macmillan.

Kittikhoun, Anoulak and Thomas G. Weiss. 2011. 'The Myth of Scholarly Irrelevance for the United Nations'. *International Studies Review* 13(1): 18–23.

Langmore, John and Jeremy Farrall. 2016. 'Can Elected Members Make a Difference in the UN Security Council? Australia's Experience in 2013–2014'. *Global Governance* 22(1): 59–77.

Loiselle, Marie-Eve. 2020. 'The Penholder System and the Rule of Law in the Security Council Decision-Making: Setback or Improvement'. *Leiden Journal of International Law* 33(1): 139–156.

Luck, Edward C. 2006. *The UN Security Council: Practice and Promise*. Routledge Global Institutions. Oxon: Routledge.

Luck, Edward C. 2016. 'The UN Security Council at Seventy: Ever Changing or Never Changing'. In *The UN Security Council in the 21st Century*, eds. Sebastian von Einsiedel, David M. Malone, and Bruno Stagno Ugarte. Boulder: Lynne Rienner.

Lupel, Adam and Lauri Mälksoo. 2019. *A Necessary Voice: Small States, International Law, and the UN Security Council*. International Peace Institute. Available at https://www.ipinst.org/wp-content/uploads/2019/04/1904_A-Necessary-Voice_Final.pdf.

Lynch, Colum. 2014. *Can Washington tame France's new tart-tongued ambassador?* Available at http://foreignpolicy.com/2014/09/09/can-washington-tame-frances-new-tart-tongued-ambassador/.

MacKenzie, David. 2015. 'Forever Adaptable: The United Nations System at 70'. *International Journal* 70(3): 489–498.

Martin, Ian. 2018. *In Hindsight: What's wrong with the Security Council?*: Security Council Report. Available at http://www.securitycouncilreport.org/monthly-forecast/2018-04/in_hindsight_whats_wrong_with_the_security_council.php.

Mody, Bella. 2010. *The Geopolitics of Representation in Foreign News: Explaining Darfur*. Plymouth: Lexington Books.

Nadin, Peter. 2016. *UN Security Council Reform*. Global Institution Series. Abingdon and New York: Routledge.

Niemetz, Martin Daniel. 2015. *Reforming UN Decision-Making Procedures: Promoting a deliberative system for global peace and security*. Routledge Research on the United Nations. Abingdon and New York: Routledge.

O'Neill, Barry. 1996. 'Power and Satisfaction in the United Nations Security Council'. *Journal of Conflict Resolution* 40(2): 219–237.

OCHA. 2014a. *Aide Memoire for the Consideration of Issues Pertaining to the Protection of Civilians in Armed Conflict*. Available at https://docs.unocha.org/sites/dms/Documents/aide%20memoire%202014%20-%20English.pdf.

OCHA. 2014b. *Security Council Norms and Practice on the Protection of Civilians in Armed Conflict: Analysis of Normative Developments in Security Council Resolutions 2009–2013*. Available at https://docs.unocha.org/sites/dms/Documents/Security%20Council%20Norms_Final%20version%20for%20print%2017%20June%202014.pdf.

OCHA. 2016. *Aide Memoire for the Consideration of Issues Pertaining to the Protection of Civilians in Armed Conflict*. Available at https://www.unocha.org/sites/unocha/files/Aide%20Memoire%202016%20II_0.pdf.

Pouliot, Vincent. 2021. 'The Gray Area of Institutional Change: How the Security Council Transforms Its Practices on the Fly'. *Journal of Global Security Studies* 6(3): 1–18.

Prantl, Jochen. 2005. 'Informal Groups of States and the UN Security Council'. *International Organization* 59(3): 559–592.

Ralph, Jason and Jess Gifkins. 2017. 'The Purpose of United Nations Security Council Practice: Contesting Competence Claims in the Normative Context Created by the Responsibility to Protect'. *European Journal of International Relations* 23(3): 630–653.

Ralph, Jason, Jess Gifkins, and Samuel Jarvis. 2020. 'The UK's Special Responsibilities at the United Nations: Diplomatic Practice in Normative Context'. *The British
Journal of Politics & International Relations* 22(2): 164–181.

Risse, Thomas and Kathryn Sikkink. 1999. 'The Socialization of International Human
Rights Norms into Domestic Practices: Introduction'. In *The Power of Human Rights:
International Norms and Domestic Change*, eds. Thomas Risse, Stephen C. Ropp,
and Kathryn Sikkink. Cambridge: Cambridge University Press.

Rosenthal, Gert. 2017. *Inside the United Nations: Multilateral Diplomacy up Close*.
Abingdon and New York: Routledge.

Rushton, Simon. 2008. 'The UN Secretary-General and Norm Entrepreneurship:
Boutros Boutros-Ghali and Democracy Promotion'. *Global Governance* 14(1):
95–110.

Schia, Niels Nagelhus. 2013. 'Being Part of the Parade—"Going Native" in the United
Nations Security Council'. *Political and Legal Anthropology Review* 36(1): 138–156.

Schia, Niels Nagelhus. 2017. 'Horseshoe and Catwalk: Power, Complexity, and
Consensus-Making in the United Nations Security Council'. In *Palaces of Hope:
The Anthropology of Global Organizations*, eds. Ronald Niezen and Marie Sapignoli.
Cambridge: Cambridge University Press.

Schindlmayr, Thomas. 2001. 'Obstructing the Security Council: The Use of the
Veto in the Twentieth Century'. *Journal of the History of International Law* 3(2):
218–234.

Security Council Report. 2012. *Council Consultations on Draft Syria Resolution*. Available at http://www.whatsinblue.org/2012/01/council-consultations-on-draft-syria-
resolution.php.

Security Council Report. 2013. *Monthly Forecast*. New York. Available at http://
www.securitycouncilreport.org/atf/cf/%7B65BFCF9B-6D27-4E9C-8CD3-
CF6E4FF96FF9%7D/2013_09_forecast.pdf.

Security Council Report. 2018a. *The Penholder System*. New York. Available at
https://www.securitycouncilreport.org/atf/cf/%7B65BFCF9B-6D27-8CD3-
CF6E4FF96FF9%7D/Penholders.pdf.

Security Council Report. 2018b. *Security Council Working Methods: Provisional
Progress*. Available at http://www.securitycouncilreport.org/atf/cf/%7B65BFCF9B-
6D27-4E9C-8CD3-CF6E4FF96FF9%7D/research_report_working_methods_
2018.pdf.

Security Council Report. 2019. 'Lead Roles within the Council in 2019: Penholders and Chairs of Subsidiary Bodies'. Accessed February. Available at https://
www.securitycouncilreport.org/monthly-forecast/2019-02/lead-roles-within-the-
council-in-2019-penholders-and-chairs-of-subsidiary-bodies.php.

Security Council Report. 2022. *2022 Chairs of Subsidiary Bodies and Penholders*. Available at https://www.securitycouncilreport.org/atf/cf/%7B65BFCF9B-6D27-4E9C-
8CD3-CF6E4FF96FF9%7D/working_methods_penholders_chairs.pdf.

Sievers, Loraine and Sam Daws. 2014. *The Procedure of the UN Security Council, 4th
ed.* Oxford: Oxford University Press.

Sievers, Loraine and Sam Daws. 2018. 'With Adoption of Three Key Documents,
Momentum Accelerates for Enhanced E10 Impact'. Update Website of 'The Procedure of the UN Security Council, 4th Edition'. Available at https://www.scprocedure.
org/chapter-3-section-3i.

Thompson, Alexander. 2010. *Channels of Power: The UN Security Council and US
Statecraft in Iraq*. New York: Cornell University Press.

Thorhallsson, Baldur. 2012. 'Small States in the UN Security Council: Means of Influence?'. *The Hague Journal of Diplomacy* 7(2): 135–160.

Traub, James. 2010. *Unwilling and Unable: The Failed Response to the Atrocities in Darfur*. New York Available at http://www.responsibilitytoprotect.org/GCR2P_UnwillingandUnableTheFailedResponsetotheAtrocitiesinDarfur.pdf.

United Nations. 1945. *Charter of the United Nations*. United Nations.

United Nations. 1982. *Provisional Rules of Procedure of the Security Council*. New York. Available at http://www.un.org/en/sc/about/rules/.

United Nations. 2004. *A More Secure World: Our Shared Responsibility. Report of the High-Level Panel on Threats, Challenges and Change*. https://www.un.org/peacebuilding/content/more-secure-world-our-shared-responsibility-%E2%80%93-report-high-level-panel-threats-challenges-and

United Nations. 2007. *5727th Meeting United Nations Security Council*. New York, 31 July, S/PV.5727.

United Nations. 2022. UN Security Council Meetings & Outcomes Tables Dag Hammarskjöld Library. Available at https://research.un.org/en/docs/sc/quick/meetings/2022.

Urquhart, Brian E. 2007. 'The Evolution of the Secretary-General'. In *Secretary or General? The UN Secretary-General in World Politics*, ed. Simon Chesterman. Cambridge: Cambridge University Press.

US Permanent Mission to the UN. 2007. *AU/UN Hybrid in Darfur: Narrowing Differences*. Wikileaks. Available at http://wikileaks.org/cable/2007/07/07USUNNEWYORK563.html#.

von Einsiedel, Sebastian, David M. Malone, and Bruno Stagno Ugarte. 2015. *The UN Security Council in an Age of Great Power Rivalry*. United Nations University. Available at https://i.unu.edu/media/cpr.unu.edu/attachment/1569/WP04_UNSCAgeofPowerRivalry.pdf.

Wallensteen, Peter and Patrik Johansson. 2004. 'Security Council Decisions in Perspective'. In *The UN Security Council: From the Cold War to the 21st Century*, ed. David M. Malone. Boulder: Lynne Rienner.

Weiss, Thomas G., Tatiana Carayannis, and Richard Jolly. 2009. 'The "Third" United Nations'. *Global Governance* 15: 123–142.

Weiss, Thomas G., David P. Forsythe, Roger A. Coate, and Kelly-Kate Pease. 2007. *The United Nations and Changing World Politics*, 5th ed. Boulder: Westview Press.

Weiss, Thomas G., David P. Forsythe, Roger A. Coate and Kelly-Kate Pease. 2017. *The United Nations and Changing World Politics*, 8th ed. Boulder: Westview Press.

Wenaweser, Christian. 2016. 'Working Methods: The Ugly Duckling of Security Council Reform'. In *The UN Security Council in the Twenty-First Century*, eds. Sebastian von Einsiedel, David M. Malone, and Bruno Stagno Ugarte. Boulder: Lynne Rienner.

Yamashita, Hikaru. 2007. 'Reading "Threats to International Peace and Security" 1946–2005'. *Diplomacy and Statecraft* 18(3): 551–572.

# 2

# Legitimation Practices

*When we study legitimacy, we are studying the behaviour of that society.*

Clark (2005: 24).

Legitimation practices moderate the effects of institutional power. Institutional power, as outlined in the previous chapter, mediates relationships within an institution, and explains how the UN Security Council is bifurcated between the permanent members and the elected members. The divisions between permanent and elected members extend beyond the formal differences in the UN Charter—where permanent members have unlimited tenure and can individually block decisions—and elected members are limited to two-year terms and cannot individually block decisions. Institutional power mediates relationships between permanent and elected members in such a way that permanent members also adopt dominant roles in agenda-setting and drafting and regularly decide matters between themselves with limited scope for input from elected members. As such, the power wielded by permanent members extends beyond the privileges granted to them by the formal rules. This power and influence is not, however, absolute; legitimation practices moderate the effects of institutional power.

Most research on legitimacy and the UN Security Council considers either the legitimacy of the Council as an institution and its decisions, or the legitimation of the Council and its decisions (Hurd 1999, 2002, 2007; Morris and Wheeler 2007). Indeed, it is common for research on international organizations in IR to not focus on decision-making processes at all (Smith 2006: 7). The usual orientation of legitimacy research is to analyse how the Security Council, or its decisions, appear to external audiences. Research in this vein arrives at judgements on how and whether the Security Council and its decisions are viewed as legitimate by various audiences (Binder and Heupel 2014; Thompson 2006, 2010; Welsh and Zaum 2013). This body of literature makes useful arguments on the legitimacy of the Security Council as an institution, and its decisions, however, it pays less attention to questions of process. Instead, the focus here is on the processual dimension. Legitimation practices both shape and enable decisions in the Security Council.

This book analyses how legitimation practices function inside the Security Council and how they shape negotiations. As such, the focus is on what happens inside the Security Council rather than perceptions of the institution or its

*Inside the UN Security Council.* Jess Gifkins, Oxford University Press. © Jess Gifkins (2023).
DOI: 10.1093/oso/9780192869029.003.0003

decisions from outside the Council. Research on Security Council practices is an emerging and critical area, as these practices have a fundamental impact on decision-making, yet are informal, with limited guidance from the UN Charter or Rules of Procedure. As Ian Hurd has argued, 'informal practices will always be worth investigating because, on the one hand, the formal rules cannot specify every contingency, and on the other hand, human creativity seems to be naturally inclined to subvert formal procedure' (1997: 140). Foregrounding these practices sheds light on how the Security Council reaches decisions and the patterns of action that are socially meaningful to participants.

The legitimation practices identified and explained in this book are internal legitimation practices—whereby drafters take patterned actions during negotiations to enhance the legitimacy of a decision. The book also identifies external legitimation practices—where seeking legitimacy shapes the actions of those in the negotiations. Traditional approaches to Security Council negotiations rely on theorizing the overarching interests of states which helps to establish the 'red lines' in negotiations but does little to explain what happens on a daily basis. As Adler-Nissen and Pouliot explain, 'While IR theories may help identify *who* pulls the strings of multilateral diplomacy, they are less useful to understand *how* strings actually get pulled' (emphasis original 2014: 890). Practice theory is particularly useful for analysing decision-making because it shifts attention onto processes. Elsewhere Adler-Nissen has argued that, 'IR scholars have ignored that diplomacy helps constitute world politics' (2015: 285). This book uses a more dynamic version of legitimacy—by analysing legitimation practices in the process of reaching decisions—to show how actors' understandings of legitimate actions and arguments shape and constrain Security Council negotiations.

Legitimation practices are socially meaningful patterns of activity that serve to enhance the legitimacy of an actor or an action.[1] They are informal by nature rather than part of the formal rules or legal requirements. Legitimation practices include activities that are reflexive and conscious—for example, the formal summit photo taken at G20 events (Hopf 2018). However, we do not necessarily have access to information about how an actor intended an action to be understood, so this book takes the actions as the object of analysis, regardless of intent, unless there are clear statements of intent (Krebs and Jackson 2007). This chapter unpacks the ways that legitimation practices shape decisions and decision-making.

The chapter proceeds in four parts. First, Security Council procedures and the advantages of a practice theory approach. Second, an overview of existing research on legitimacy and the Security Council. Third, a section that sets out three internal legitimation practices, which enhance the legitimacy of a decision. These are the drive towards unanimity, above and beyond the legal requirements, use of language that has been agreed in previous decisions, and seeking support from states

---

[1] This definition draws on the definition of practices put forward by Adler and Pouliot (2011a: 4).

who reside in the region which is under discussion. These internal legitimation practices facilitate agreement and enhance the legitimacy of the decision under discussion. Fourth, a section which sets out two external legitimation practices whereby states take actions to enhance their own legitimacy in the eyes of key domestic or international audiences. These are the use of consistent arguments and taking action to be seen to be 'doing something', even when the actions cannot be implemented. These external legitimation practices enhance the legitimacy of the state in relation to key audiences. This chapter establishes the utility of focusing on legitimation practices within negotiations and the second part of the book uses this framework to analyse the Security Council's negotiations on Darfur.

## Security Council Procedures

To understand Security Council decision-making, you could look to the formal rules and structures that govern the Council; however, the rules and structures offer little insight into what happens each day. The UN Charter is deliberately broad and flexible with the capacity to adapt to new threats (Luck 2008). The Charter outlines the mandate of the UN Security Council but offers little guidance on how to interpret the mandate or on day-to-day operations. Aptly described by Franck, if a Martian attempted to understand how states interact by reading the UN Charter, they would be very disappointed (1990: 21–22). Even the parameters of the Security Council's mandate—as set out in the UN Charter—have evolved considerably over time, as the Security Council has interpreted and reinterpreted what constitutes a threat to international peace and security (Yamashita 2007). One of the first tasks for the UN Security Council was to design its own working methods, however, there was, and continues to be, reluctance on committing to fixed working methods, so the Provisional Rules of Procedure have remained 'provisional' since they were established in 1946. As an interviewee explained to Schia, the Rules of Procedure are not particularly relevant to negotiations because 'the UNSC decides by itself, a [sic] 100 percent, its own rules of procedure and how it chooses to work' (2017: 68). As such, the formal rules—both the UN Charter and the Provisional Rules of Procedure—offer little guidance on the day-to-day process of decision-making.

If the formal rules offer little insight into decision-making, can we turn to collated repositories of working methods to understand how negotiations occur? In recent years, there have been attempts by Security Council members to collate Council practices into a recurring series of notes, instigated by Japan, collected under the title of the first iteration: Note 507. However, there is a key limitation in looking to these notes to understand the practice of decision-making because they serve partly to document existing practices and partly in 'aspirational terms' describing how (some) Security Council members would like decision-making to operate (Pouliot 2021; Security Council Report 2018b: 10). Note 507s do not distinguish between sections that are descriptive and sections that are aspirational. For example, the practice of penholding—whereby one or more state takes political ownership for drafting decisions on a given topic—is discussed in the 2017 Note 507 where it is stressed that 'Any member of the Security Council may be a penholder' (United Nations Security Council 2017). This has also been stressed in previous Note 507s, yet in practice, P3 members (the UK, the US, and France) maintain their stranglehold on the practice of penholding. In 2022 for example, the P3 were penholders for 73 per cent of the topics before the Security Council (for topics with fixed penholders, across both conflicts and thematic areas) (Security Council Report 2022). Given that there is a significant gap between the formally documented working methods and the day-to-day operation of the Council, it is insufficient to analyse the rules—we also need to analyse informal practices. Analysing informal practices helps to explain the distribution of power and influence within negotiations and how practices shape what is possible in negotiations.

## Legitimacy and the UN Security Council

Pinning down legitimacy is difficult. Legitimacy is regularly described in IR in ambiguous terms as a 'slippery concept' (Frost 2013; Hurrell 2005), as a 'conceptual quagmire' (Kratochwil 2006: 303), and as 'an opaque and elusive concept' (Steffek 2003: 251). These are apt terms for a concept that is intersubjective, is only meaningful in relation to a given time and place, and connects to other pivotal concepts such as power, law and morality (Williams 2013). As such, the concept of legitimacy is most useful in relation to a specific setting and group of actors. It is not that certain actions are legitimate across time and place; rather legitimacy is 'quite literally meaningless outside of a particular historical context' because on its own 'legitimacy has no DNA' (Hurrell 2005: 29). This means that legitimacy has no inherent content but manifests in different ways in different settings. The slipperiness of legitimacy as an abstract concept, combined with its intrinsic relationship to a specific time, place, and society makes it logical to ground a study of legitimacy in relation to the activities of a group at a specific time and the group's shared understandings of what constitutes appropriate practice.

Unsurprisingly, the legitimacy of IOs and the legitimation of IOs has become a topic of considerable interest over recent decades, particularly as the remits of IOs have expanded into more areas. Recognizing that IOs often have limited power with which to enforce their decisions, the key focus within this literature is how IOs gain, lose, maintain, and justify their legitimacy (Halliday, Block-Lieb and Carruthers 2010; Tallberg and Zürn 2019; Zaum 2013). There is also a focus on how to assess the legitimacy of IOs from the vantage point of specific

audiences (Buchanan and Keohane 2006; Chapman 2009). Debates on the legitimacy of IOs also centre on questions relating to the use of force and why powerful
states use IOs to legitimate decisions on the use of force, with the understanding that IOs play a significant role for both signalling and gatekeeping (Chapman
and Reiter 2004; Coleman 2007; Finnemore 2003; Thompson 2006, 2010; Voeten
2001, 2005). The common themes among literature on legitimacy and IOs is the
legitimacy of IOs as actors (where this legitimacy comes from, and how to measure
it) and the legitimacy of decisions made by IOs, particularly when the decisions
relate to contentious issues like the use of force, rather than legitimacy within IO
decision-making.

The dominant conceptualization of legitimacy in Security Council research
focuses on rules and institutions over more relational versions of legitimacy. There
are a wide range of approaches to legitimacy, from Scharpf's input and output
legitimacy, to Applbaum's procedural and substantive legitimacy, and Steffek and
Clark each draw on Max Weber's distinction between normative and empirical
approaches to legitimacy (Applbaum 2000; Clark 2005: 18; Scharpf 1999; Steffek
2003: 252). However, the most commonly used definitions of legitimacy in the
study of the Security Council draw from foundational work by Franck. He argues
that legitimacy is 'a property of a rule or rule-making institution which itself exerts
a pull toward compliance on those addressed normatively because those addressed
believe that the rule or institution has come into being and operates in accor-
dance with generally accepted principles of right process' (1990: 24). This focus
on the legitimacy (or otherwise) of rules and institutions is evident in subsequent
research on the Security Council. For example, the most comprehensive analysis
of legitimacy in relation to the Security Council, by Ian Hurd, draws directly from
Franck and defines legitimacy as 'an actor's normative belief that a rule or institution ought to be obeyed' (2007: 7). This definitional framing has implications for
the orientation of research on the Security Council and legitimacy.

Following from this, the standard approach to analysing the relationship
between legitimacy and the Security Council is to consider the legitimacy of
the Council as an institution or the legitimacy of its decisions. These avenues
of enquiry lead to questions around the legitimation effects of Security Council
decisions (and therefore why powerful states would use the Council) (Chapman 2009; Chapman and Reiter 2004; Matsumura and Tago 2019; Thompson
2006, 2010); and the relationship between legitimacy and Security Council reform
(Chesterman 2006; Hurd 2008; Niemetz 2015; Welsh and Zaum 2013). There
is also abundant literature on the legitimacy (or otherwise) of individual decisions made by the Security Council (see for example Groom and Morphet 2006;
Morris and Wheeler 2007; Ralph and Gallagher 2015; Weiss 2014). This is unsurprising, because a common starting point for analyses of the Security Council is
the tension between the power wielded by the Council and its limited capacity
to enforce its decisions (see for example Hurd 2002). Coleman, however, shows

how International Relations has uncritically adopted a model of legitimacy from Political Theory that overly focuses on rules and institutions (2007). Broadening legitimacy beyond rules and institutions allows for a more relational version of legitimacy.

A more encompassing version of legitimacy goes beyond the focus on the legitimacy of rules or institutions. The idea that legitimacy relates to the correctness or rightfulness of a rule or institution hinges on ideas about who has the right to govern and draw from domestic politics (Coleman 2007: 21; Visoka 2020: 46). These democratic approaches to legitimacy are overly static and miss out on the relational dimension of legitimacy (Visoka 2020). A broad review of definitions of legitimacy found commonalities among definitions whereby legitimacy is: the construction of a social object consistent with shared beliefs and norms; it is a collective process; it relates to the perception of consensus; and it includes both a cognitive dimension and a prescriptive dimension (Johnson, Dowd and Ridgeway 2006: 57). These features stress the social and intersubjective nature of legitimacy. In building up a more relational vision of legitimacy in relation to the Security Council, Coleman argues for adding actors and actions to our analysis, in addition to rules and institutions (2007: 21). I also add arguments to this list, as the social assessment of arguments is central to the deliberative function of the Security Council.

The approach taken in this book has some similarities with the idea of self-legitimation practices, used by Barker, von Billerbeck, and others (Barker 2001; von Billerbeck 2019, 2020). Their work on self-legitimation practices shifts the focus from external audiences to internal audiences within IOs, whereby specific actions are taken to reinforce the belief by staff that the work undertaken by an institution, and its mission, are legitimate, or the steps a ruler takes to legitimate their actions. Self-legitimation practices have some similarities with the legitimation practices developed in this book in that both analyse legitimation within an organization. However, where von Billerbeck analyses the ways that employees legitimate their organization to themselves, the approach taken in this book is to analyse how legitimation affects negotiations; it is analysing how the desire to be seen as legitimate shapes the actions and arguments inside the Security Council. The object of focus is not the Security Council and its legitimacy as an institution, rather how legitimation practices constrain and enable decisions and shape what is and is not possible. Unlike the approaches on self-legitimation, this book considers audiences which are both internal and external to an IO.

Legitimacy, and perceptions thereof, go beyond the formality of rules and institutions and types of legitimacy that are relevant in the Security Council include actors, actions, and arguments. Coleman put an alternative and more dynamic conception of legitimacy forward. She defined legitimacy as 'a social status that can adhere to an actor or an action: it involves being recognised as good, proper, or commendable by a group of others' (2007: 20). This definition is better suited to analysing legitimacy in negotiations because it starts from an actor or action rather than starting from an *ex post facto* assessment of a rule or institution. The definition includes an audience (which can evaluate legitimacy) and by framing it as a social status we are reminded that legitimacy is something that actors seek. As Coleman explains, her definition is 'other-regarding rather than self-regarding; states need not inherently want to act as good citizens, they simply want to be seen as such' (2007: 38). She recognizes that legitimacy is only one motivating factor for states, however, and shows that states 'value their standing in international society as an end in itself, not simply as a means to other ends, and they chafe as international opprobrium itself, not only at its possible material consequences' (2007: 39). While Coleman's definition assumes that legitimacy can adhere to an actor or an action, it can also adhere to an argument, which is relevant in the context of Security Council negotiations where varied arguments are put forward in negotiations and some arguments come to be seen as legitimate while others do not. Key points, then, are that legitimacy can adhere to an actor, action, or argument; that it is intersubjective; that it involves audience judgements; and that it is a social status and one that actors seek. This way of understanding legitimacy is more dynamic and is better suited to analysing legitimacy within the process of negotiation.

Status is central to this social understanding of legitimacy and to the conduct of diplomacy. Status refers to 'an individual's standing in the hierarchy of a group based on criteria such as prestige, honour, and deference' (Johnston 2001: 500). The quest for status and hierarchy within the pecking order is integral to multilateralism (Pouliot 2016). For example, in recent years there has been increased debate in over the status of the UK in international politics, particularly in relation to Brexit (Gifkins, Jarvis, and Ralph 2019; Morris 2011; Oliver and Williams 2016; Ralph, Gifkins, and Jarvis 2020). Schia and Sending explain that 'status is sought and accorded in and through concrete practical encounters, mediated and structured by the rules that define diplomacy' (2015: 75). Status is so central to the practice of multilateralism that they conclude, 'The primary driver of diplomats' behaviour in the UN is the quest for status' (2015: 82). There is a close relationship between legitimacy as a social status and the practice of diplomacy.

## Practice Theory and International Relations

Practices and legitimacy are related because practices are the actions that are recognized by insiders as the normal and correct way of doing things. Practices are 'competent performances. More precisely, practices are socially meaningful patterns of action, which, in being performed more or less competently, simultaneously embody, act out, and possibly reify background knowledge and discourse in and on the material world' (Adler and Pouliot 2011a: 4, 2011b). The socially meaningful nature of practices means that they necessarily make sense to other

practitioners. Practices make sense within a given field, be that medicine, education, or diplomacy; that is, participants recognize practices as the appropriate ways
of doing things within a given field. Performances that are recognized as competent by fellow practitioners are those that are legitimated at a given time and within
a given setting. As such, there is a strong relationship between practices and legitimacy. This does not necessarily mean that practices have normative legitimacy
(Ralph and Gifkins 2017). Indeed, a practice could be viewed as normatively rep-
rehensible but have continued practice within a given setting at a given time. This
is to say that the legitimation of a practice is not a normative judgement, but an
acknowledgement within a community of practice that a pattern of behaviour is
normal and accepted at a given time and place.

There has been extensive debate within International Relations over the extent
to which practices are reflexive—that is, enacted with conscious awareness by
practitioners—or not, to the point where Hopf argues that IR has minimized
the potential for reflexive practices, which earlier social theorists understood as
possible (2018). An early, and influential, intervention in this debate within IR
was made by Pouliot, who initially framed practices as inarticulate and thought-
less (2008). Subsequent scholarship in IR more often presumes the potential,
at least, for practitioners to think reflexively about practices (Adler and Pouliot
2011a; Bode and Karlsrud 2019; Bueger and Gadinger 2018; Hopf 2018). For
the purposes of this book, it is possible, although not essential, for practices to
be enacted consciously, deliberately, or even strategically. Unlike constructivism,
which presumes a process of socialization into norms, and internalization of
norms, practices are performances that presume an audience, so the question of
how conscious the practice is, is less central because we can establish evidence of
the practice without needing to rely on a conscious intention behind the performance. As Pouliot later argued practice theory 'stops short of trying to get inside
people's heads ... After all, practices are public objects, that is, they are performed
in front of an audience' (2016: 18). More recent work by Pouliot offers the concept of a semi-formal 'grey area' which distinguishes between secondary rules (i.e.,
rules about rules), which are formal, such as charters; informal rules, such as cus-
tom; and semi-formal rules, which are self-monitoring practices (2021). Pouliot
elaborates, 'For practice theorists, the grey area offers a useful reminder that even
when practices change in and through practices, reflexivity may still play a role'
(2021: 5). Much of the day-to-day practices of the UN Security Council fit within
the semi-formal grey area, meaning that practices may be enacted reflexively, however, the performance of the practice is more relevant for the purposes of this
book.

Another key debate about practice theory within IR, and beyond, is the specific
relationship between practices, and other repetitive actions such as habits and routines. There are some key differences between these concepts, however. Habits can
be understood as unreflexive and automatic, and while routines are repetitive they

do not contain the same potential for innovation that practices do (Bueger 2014; Hopf 2010). Practices are repetitive actions; however, 'they are also permanently displacing and shifting patterns' (Bueger 2014). To put this another way, practices contain the potential for change (Hopf 2018). Unlike habits or routines, practices can be assessed by a knowledgeable audience as a correct or incorrect performance or as a more or less competent performance (Adler-Nissen and Pouliot 2014; Neumann 2002). In this sense, practices are socially meaningful (Adler and Pouliot 2011a, 2011b). Practices, then, are repetitive like habits but are assessed by an audience and contain the potential for change.

The example of penholding, whereby a state takes 'political ownership' for a given issue area in the Security Council, helps to illustrate this. In theory, any state can be penholder, a point emphasized in multiple Note 507s, which are compendiums of Security Council working methods (Security Council Report 2018a). However, in practice, the UK, the US, and France have the vast majority of penholding roles (Security Council Report 2022). While some elected members find this problematic, there can be contestation over something that is still ultimately understood as legitimate because legitimacy is a collective assessment (Johnson, Dowd and Ridgeway 2006). The political ownership implicit in penholding is recognized as socially meaningful by practitioners, and despite the frustrations of elected members who find it limits their input into drafting, it is rarely challenged (Ralph and Gifkins 2017). Practices contain the potential for change, however, and in 2013 when the Security Council was deadlocked over negotiations on Syria, Australia, Luxemburg, and Jordan, 'picked up the pen' on the Syrian humanitarian track and were successful in securing a unanimous resolution, despite the 'scepticism' of the P5 (Ralph and Gifkins 2017: 646–647). We can ascertain that the performance of this group of elected members has been assessed as competent by fellow Security Council members, as an evolving group of elected members have maintained control over penholding for the Syrian humanitarian track (Security Council Report 2020). The example of the practice of penholding illustrates key features of practices, such as their socially meaningful nature, the potential for change, and the relationship with competence, which are different to habits and routines.

Another point that needs clarifying is the relationship between states and diplomats, from a practice theory perspective. As noted in the Introduction, one of the advantages of practice theory is that it refocuses attention onto diplomats and their daily interactions, while much of IR focuses on 'states', in the abstract, rather than the activities of diplomats. Adler-Nissen goes as far as to say that diplomats find mainstream International Relations theory 'strange' as it 'does not capture the bodily experience of being a diplomat' (2015: 285). While refocusing on the day-to-day activities of international politics is important, a hard line between states and diplomats is also unadvisable. This is an issue that Pouliot grapples with in his 2016 book, demonstrating that, rather than diplomats acting as a simple 'conveyor belt' to represent their instructions from capital, some diplomats are actively engaged in the process of policy formation, and he instead argues for the constitutive power of diplomacy (2016: 34). States vary on how much discretion or leeway they give to their diplomats. For example, British and French diplomats are considered to have a high degree of leeway (Gifkins, Jarvis, and Ralph 2019: 1356; Pouliot 2016: 202). There are also examples, albeit rare ones, where a diplomat acts against their instructions from capital (Schia 2013). Regardless of how much leeway individual diplomats have, or whether they are following their instructions, diplomats are understood as the representatives of states, and we can learn something from their actions about state behaviour. As such, analysing the practices of diplomats offers insight about the behaviour of states.

## Legitimation Practices

Legitimation practices are a synthesis between legitimation and practice theory. They are repeated patterns of behaviour that serve to reinforce the legitimacy of an actor or a decision. Legitimation practices are common, both in domestic politics and within multilateralism more broadly. At the domestic level, legitimation practices that enhance the legitimacy of a decision would include launching a policy with great fanfare and media engagement, or securing broad stakeholder buy-in for such a policy. A domestic legitimation practice that reinforces the legitimacy of an actor would include the pageantry of coronation for a new monarch. At the multilateral or bilateral level, legitimation practices include prioritizing unanimity to shore up the legitimacy of a decision, or heads of state congratulating international counterparts on their election. These actions are common, repeated, recognizable, and meaningful to participants, and reinforce the legitimacy of an actor or a decision.

There is extensive literature on legitimacy in relation to IOs and rapidly growing interest in practice theory and IOs; however, the synthesis is significant because it foregrounds the repeated and socially meaningful nature of legitimation activities, and shows how these are patterned but also contain the potential for change. The synthesis of legitimation and practices is also better suited to the analysis of decision-making, and the implications of these practices on negotiations, beyond the typical focus on the legitimacy of a decision or an institution. In recent years, two articles have referred to legitimation practices, without, however, developing the concept in any depth (Spandler 2020; Tallberg and Zürn 2019). They recognize the advantages that this concept brings in terms of moving away from a static understanding of legitimation, but these articles still focus on the legitimation of IOs rather than decision-making. As such, the sustained development of legitimation practices, as set out in this book, represents an innovation on the analysis of decision-making in IOs.

In conceptualizing legitimation practices, I have divided them into two categories; internal legitimation practices and external legitimation practices, each discussed in turn in the following sections. These are internal or external from the vantage point of the Security Council, whereby internal legitimation practices relate to the legitimacy of a decision and external legitimation practices relate to the legitimacy of states to relevant domestic and international audiences. While both practices relate to legitimation, the object of legitimation is different: a Security Council decision or an actor in the negotiations. Legitimation practices could be used strategically, for example, garnering regional support for a draft to help mitigate dissent; however, this could also occur in a taken-for-granted manner, with the drafter engaging with regional actors as part of the normal process of negotiation. As previously discussed, practice theory does not require an assessment of conscious intentions because practices are public objects with an audience (Pouliot 2016). There is a spectrum here, however, whereby internal legitimation practices such as unanimity or repetition of language are at the more taken-for-granted end of the spectrum, and external legitimation practices such as 'doing something' are likely to be more deliberate because they connect with domestic politics and questions of international prestige. Internal and external legitimation practices both have an important and underrated impact on the process and outcome of Security Council negotiations.

Within Security Council negotiations, there is extensive evidence of legitimation practices. By legitimation practices, I mean patterned actions that are recognizable to relevant audiences as conforming to shared understandings of appropriate behaviour to gain a reward in social standing for themselves or for the decision. These are common practices within Security Council negotiations, and they shape both the process and outcome of these negotiations. For example, Adler-Nissen and Pouliot describe how P3 states encouraged and amplified the voices of specific elected members to increase the legitimacy and support for their plans of a no-fly zone over Libya (2014). This included amplifying Bosnia's voice on the utility of no-fly zones where it could speak from experience and asking the defected Libyan ambassador to request that the Security Council refer the situation to the International Criminal Court for investigation. Each of these actions served to increase the social standing of the decision. Legitimation practices are common to Security Council negotiations.

## Internal Legitimation Practices

There are two types of legitimation practices identified in this book: internal legitimation practices, which refer to legitimation of the decision; and external legitimation practices, which refer to legitimation of the actor. Internal legitimation practices are routine actions taken during negotiations to enhance the

legitimacy of a decision, of which there are three key types. First, prioritizing unanimity in making decisions, despite the legal requirement of only nine affirmative
votes (and no vetoes). Security Council members prioritize unanimity beyond the
legal requirements. Second, repetition of language and phrases within Security
Council decisions because states realize that language that has been 'previously
agreed' has an additional weight of legitimacy that can enhance support and ease
dissent (Gifkins 2016b; Werner 2017). Third, states that are geographically connected (or sometimes culturally connected) to the situation under discussion are
highly valued by drafters for the legitimacy that their support brings to a decision. This can be individual states inside the Security Council, or relevant regional
groups outside the Security Council which have been increasingly seen to play a
'gatekeeper' role within negotiations (Bellamy and Williams 2011).[2] These three
internal legitimation practices are all routine actions taken by drafters which
enhance the legitimacy of a decision, for example, if a decision is unanimous,
draws on language which has been previously agreed in other decisions, and has
strong support from states within the region under discussion, it will have a high
degree of legitimacy. Each of these three internal legitimation practices are set out
in turn below.

## Unanimity

Analysing the way that legitimacy shapes the process of decision-making brings
into focus the drive towards unanimous resolutions. As the former Department
of Peacekeeping Operations stated in its Capstone Report, 'anything other than
unanimous Security Council backing can be a serious handicap', despite the fact
that this makes no legal difference to a resolution (DPKO 2008: 50). Legitimacy,
as argued by Whalan, 'makes the exercise of power easier, less costly, more effective and more resilient' (cited in Visoka 2020: 49). Where there is a trade-off to
be made, Wood argues that 'the drafting of Security Council resolutions is often
a highly politicized process, with a desire for consensus being more important
than legal precision' (2016: 12). As some have highlighted, most decisions made
by the Security Council since the end of the Cold War have been unanimous,
far above and beyond the legal requirements of nine affirmative votes (and no
vetoes) (Dunne and Gifkins 2011; Hulton 2004; Williams 2013). However, there
has been a lack of research on the way that prioritizing unanimous voting affects
negotiations. This is surprising, given that consensus decision-making has been
highlighted as one of the two most significant changes in the Security Council

---

[2] Another minor internal legitimation practice is co-sponsoring resolutions, which holds no legal
consequence but enhances the legitimacy of a decision. Drafters seek co-sponsorship as a 'final stamp'
of legitimacy for a decision at the stage of voting, however, this does not appear to substantively shape
the process of negotiation in most cases.

since the end of the Cold War (von Einsiedel, Malone and Stagno Ugarte 2015: 3). In fact, the former Permanent Representative for Guatemala, Gert Rosenthal, describes 'consensus-building' as 'virtually a synonym for decision-making at the United Nations' (2017: xiii). The preference for unity is so strong that between 2000 and 2021 87 per cent of drafts put to a vote have not only passed but also passed unanimously.[3] The drive towards unanimity is a key aspect of the Security Council's internal legitimation practices, which shapes the process of reaching decisions.

In addition to the observable pattern of unanimous voting, there are also statements made after a vote, which support the idea that Security Council members privilege unanimity. It is common for states to express regret or disappointment when the lack of consensus blocks a resolution, either due to a negative vote from a P5 member, or failure to reach the required majority.[4] In these cases, the non-unanimous vote changes the outcome of the draft, so it is unsurprising that some states would express regret or dissatisfaction when this occurs. However, when a resolution passes, but without unity, states still often express disappointment, even though the abstentions do not change the legal outcome.[5] There are also indications that states find it difficult to abstain on a resolution that would otherwise be unanimous (Adler-Nissen and Pouliot 2014: 900 and 902; Schia 2013). Pouliot has explained that 'multilateralism creates strong pressures of conformism' (2016: 125). As an example, when the Intervention Brigade was mandated in the Democratic Republic of the Congo, the Security Council vote was unanimous; however, Argentina, Guatemala, and Pakistan each stated at the time of the vote that they had serious and ongoing concerns with the resolution which had not been addressed (United Nations 2013). Guatemala described these as 'concerns that caused us to waver in joining the consensus' (United Nations 2013: 4). Despite these ongoing concerns at the time of the vote, all states voted for the Intervention Brigade. These examples and patterns of behaviour suggest that the preference for unanimous voting is so strong in the Security Council that there is some 'compliance pull' towards affirmative voting, to use Thomas Franck's language (1990). Considerations of legitimacy, and the legitimacy that comes from unanimous voting, shape Security Council negotiations via encouraging affirmative voting and discouraging divisions in voting. Outcomes are seen as more legitimate if they are

---

[3] Data compiled by the author from UN (2022). It is worth noting that while the average remains strong, the level of unanimity has decreased in recent years, with an average of only 74 per cent unanimous from 2018 to 2020.

[4] For example, see statements by Guatemala, Russia, the UK, France, Morocco, Australia, Togo, Rwanda, and China on draft S/2013/660. Another example can be found in statements by the UK, Germany, Pakistan, India, Portugal, Guatemala, the US, South Africa, Morocco, China, and Colombia on draft S/2012/538.

[5] See for example, statements by China and Russia on resolution 2182; statement by the UK on resolution 2081; statements by China, Azerbaijan, Russia, France, Togo, the UK, Pakistan, India, and Germany on resolution 2068; and statement by Pakistan on resolution 2058.

unanimous, and indeed lack of unanimity can be perceived as a weakness of the resolution. A focus on legitimation practices brings normal patterned actions like unanimity into the foreground.

It is worth noting, however, that while there is a general drive towards unanimity in Security Council negotiations, and this influenced some stages of the Darfur negotiations, Darfur was one of the most contentious matters before the Council at the time and this is reflected in voting patterns as shown in Figure 2. For example, between 2004 and 2007—the years when the Security Council applied key tools to the conflict in Darfur for the first time—there were twenty-one resolutions across all topics voted on by the Security Council that had some form of division.[6] These resolutions either did not pass because they were vetoed or passed but with abstentions or negative votes from elected members. Of the twenty-one, six were on Darfur, which gives some indication of how divisive the situation in Darfur was in comparison to other topics under discussion during the same period. The drive towards unanimity can be evidenced more fully via a meta-study of Security Council negotiations, rather than an individual case study, however, there were still some effects of the drive towards unanimity in the case of Darfur, as drafters sought (but did not always achieve) the social status of legitimacy that comes from unanimous decisions.

## Language

Repetition of language in Security Council decisions is another key legitimation practice. Citing previous decisions in a current draft enables the Security Council to draw on the legitimacy of earlier decisions as well as reinforce its significance to the matter at hand (Werner 2017). Barnett explained that to operate within the Security Council he had to learn to speak and write the language in a way that made sense to insiders (1997). Repetition is a key part of 'speaking the language' and is a common feature of foreign policy writing. Indeed, Neumann argues that because repetition is a form of reaffirmation, policies need to be repeated otherwise they are weakened (2007). Within the Security Council, it is not only the political imperative of reaffirmation that drives repetition, but also that once states have agreed to a certain phrasing it is harder for it to be rejected in the future, so 'agreed language' becomes the bedrock of future negotiations. As Niemann explains, 'agreed language creates entanglements and expectations' (2022: 265). This can be seen in the Security Council's use of language on R2P, which was highly contentious the first time it was applied to a conflict—Darfur—but became routine and commonplace afterwards (Gifkins 2016a). Across different policy issues, OCHA produces an 'Aide Memoire for the Consideration of Issues Pertaining to the Protection of

---

[6] Data compiled by the author from UN (2022).

Civilians in Armed Conflict' which includes a section on 'agreed language' (OCHA 2016). Agreed language can be, and is, copied and pasted into new drafts to ease negotiation (Adler-Nissen and Drieschova 2019; Howard and Dayal 2017). Repetition of language then becomes integral to both reaffirmation of policies and as a strategic tool to assist in the drafting process via leveraging 'agreed language' as a starting point in negotiations. Agreed language helps to facilitate agreement and helps to legitimate a resolution.

## Regional Relevance

Drafters use the legitimacy of regionally relevant actors—whether they are states or regional groups—to enhance the legitimacy of a decision. As discussed Chapter 1, elected states can have additional influence in negotiations when they are from the region under discussion. Drafters use this strategically to demonstrate 'regional support' for a decision. This can be with the support of individual states, for example when the French Permanent Representative to the UN sought support for a resolution on Children and Armed Conflict he 'brought elected member Benin on board "to demonstrate West-African involvement"' (Bode 2018: 307). States can also bring connections to relevant regional groups. For example, during negotiations towards a resolution on Libya, Lebanon was encouraged by drafters to speak, as a member of the Arab League that supported NATO's intervention, and to help coordinate between the Security Council and the Arab League (Adler-Nissen and Pouliot 2014: 899). The significance of regional groups is growing, and they have shown an emerging role as 'gatekeepers' in enabling some Security Council activities (Bellamy and Williams 2011). Drafters recognize the strategic value of support from the region under discussion, both in the form of individual state support and formal support from the relevant regional group, and seek this as a normal part of negotiations.

## External Legitimation Practices

In addition to the internal legitimation practices which enhance the legitimacy of a decision, this book also identifies external legitimation practices which are patterned actions taken during negotiations that serve to enhance the legitimacy of an actor. External legitimation practices take two key forms. First, the arguments that states make in the Security Council need to appear legitimate to negotiating partners and to domestic and international audiences. Arguments need to contain a degree of consistency and to draw on rationales other than simple self-interest, which can then constrain the range of options subsequently available to actors (Booth Walling 2013; Elster 1995; Johnstone 2003; Krebs and Jackson

2007). Second, states want to be seen to be 'doing something', even when there is not sufficient agreement to pass a decision which is implementable. The legitimacy that comes from being seen to take action can incentivize states to pass a decision, even if that decision subsequently cannot be implemented, as there is greater international attention on decisions than there is on implementation. These two external legitimation practices are evident in Security Council negotiations as states seek legitimacy in the eyes of audiences that are important to them. There is a relationship between how high profile a negotiation is and how much impact external legitimation practices are likely to have. As Coleman highlights, 'The more likely an activity is to command international attention, the greater are both the risk of incurring international opprobrium and the chance of garnering international prestige' (2007: 38). Both of these external legitimation practices are set out in turn below.

## Arguments

States are constrained in the arguments they can make in the Security Council through consideration of audiences that are important to them. There is necessarily a close connection between legitimacy and audiences (Coleman 2007; Welsh and Zaum 2013; Williams 2013). Legitimacy does not exist in the abstract but in perceptions of an actor, action, or argument by a given audience. There is also a strong relationship between international institutions and public opinion (Chapman 2007, 2009; Chapman and Reiter 2004; Thompson 2006, 2010). The focus here, however, is not how domestic audiences receive arguments made in the Security Council, but how concerns around perceptions by audiences—domestic or international—enable and constrain the types of arguments states make within this body. In the case of Darfur, state's consideration of how important audiences might receive a particular argument can explain key turning points. For example, the US and the UK shifted from outright opposition to any discussion of Darfur in the Security Council to championing Darfur as an agenda item. The US also enabled the referral of Darfur to the International Criminal Court, via an abstention, despite years of opposition to the court. In addition, China shifted from actively shielding Sudan from UN peacekeeping to lobbying Khartoum to accept UN peacekeepers. Concerns around the legitimacy of arguments were central to each of these shifts, as is shown in subsequent chapters.

States strive to construct arguments that will appear legitimate to important audiences. For example, the government of Sudan did not assert that it had the right to kill their own people in Darfur, instead it argued that attacks on civilians had been fabricated and that the number of people killed had been exaggerated (see for example Natsios 2012: 149). More specifically, the government of Sudan has regularly used arguments about conspiracy or neocolonialism when talking about Western criticism of violence against civilians in an attempt to find arguments which will be regarded as legitimate by some audiences (Washburne 2009). States use arguments which will appeal more to audiences than raw self-interest would, which means they need to take other perspectives into account, which has been called the 'civilizing force of hypocrisy' (Elster 1995, 1998). As Pouliot explains, 'the pure and naked defence of national interest poses problems in multilateral diplomacy' (2016: 123–124). Once states have made a particular argument, there are then incentives to maintain this argument to appear consistent (Johnstone 2003). The advantages of making arguments that are not identical to the state's interests, combined with the advantages of consistency, can lead to some constraints for states. As Booth Walling argues, 'Whether made sincerely or strategically Council arguments have meaning and influence independent of the speaker' (2013: 245). The risks here for states have been described as 'rhetorical entrapment', meaning that once a state has made an argument, it is harder for them to then make an opposing argument (Schimmelfennig 2001). This does not mean that states always act consistently, but it does suggest that there are some constraints for states in presenting consistent arguments. There are also incentives towards adopting mainstream positions in multilateral diplomacy (Pouliot 2016: 206). States can be constrained in the arguments they make within the Security Council via consideration of audiences that matter to them, particularly around the consistency of current arguments with previous arguments.

## 'Doing Something'

The final external legitimation practice is taking action to be seen to be taking action, even when the decision will not have an impact on the ground, or even be implemented. The idea of 'doing something' appears in literature on sanctions occasionally. For example, Tierney (2005: 647) and Elliott (2009-10: 87–88) both discuss sanctions as a tool that can be used to give the appearance of 'doing something', and as a means of appealing to domestic audiences. It also appears in literature on peacekeeping and R2P (Bode and Karlsrud 2019; Guéhenno 2015; Lanz 2020). In this book, the idea of 'doing something' applies more broadly to decisions which are made to appease specific audiences, where there are public statements available that participants knew the decision could not be implemented or would have little to no impact even if implemented. When the Cold War ended there was a sudden escalation in the activism of the Security Council and the number of resolutions it generated per year. For example, between 1946 and 1989 it passed 646 resolutions, an average of 15 per year, whereas between 1989 and 2014 it passed 1549 resolutions, an average of 62 per year (Wallensteen and Johansson 2016: 28–29). This growing activism and expanded mandate have led to the valorizing of action, even when the actions authorized are not implementable. Colum

Lynch, key UN analyst, listed his 'ten worst Security Council resolutions ever', which includes a number of resolutions which could not be implemented (2010). This suggests that the drive to 'do something' can prioritize creating a decision over creating a meaningful decision.

In the case of Darfur, intense negotiations over sanctions between 2004 and 2006 created an arms embargo on Darfur (but not all of Sudan) which was very difficult to enforce, and targeted sanctions against only four individuals (Dörfler 2019). Similarly, six months of fraught negotiations towards UN peacekeeping led to a resolution which 'invited the consent' of the government of Sudan, despite Khartoum being very clear before the resolution passed that its consent would not be forthcoming. The drive towards legitimacy may result in a compromise which papers over fundamental disagreements between Security Council members in the name of 'doing something', whether this can be implemented or not. As Mills argues, 'there was very great pressure to "do something" in Darfur' (2009: 548).

Alongside the drive towards 'doing something' there is a logic of 'ratcheting' which exists inside the Security Council. Ratcheting refers to a series of actions which progress through a linear process of agenda-setting, expressing concern, initial non-military measures, intensifying non-military measures, through to military enforcement measures (Gehring and Dörfler 2019: 132). As such, once a matter is under consideration by the Security Council, there can be an internal logic towards 'doing something', which equates to making a series of decisions to use escalating measures to address the crises. The Security Council also has a tendency to 'vote and forget', with less attention paid to implementation (Sievers and Daws 2019). The effectiveness of a decision can be sacrificed within the decision-making process due to the drive to 'do something' and the ratcheting logic of Security Council negotiations.

Another aspect of the drive towards 'doing something' is that Security Council members can find text-based compromises which simply paper over real and ongoing political divisions. Ambiguity in language in resolutions can actually be intentional in the sense that it enables states to agree to a decision while maintaining divergent perspectives on what has been agreed to (Byers 2004). Lengthy periods of negotiation can encourage ambiguities as part of political compromise, rather than clarity (Wood 2016). This was evident in the statements given after resolution 1556 on Darfur. After the resolution, the US stated that 'The resolution anticipates sanctions against the Government of Sudan if the regular monthly reporting cycle reveals a lack of compliance' (United Nations 2004: 4). And Russia stated, alluding to sanctions, that 'of fundamental importance is the fact that the resolution does not foresee possible further Security Council action with regard to Darfur' (United Nations 2004: 7). In this example, states found language that they could accept which enabled them to hold different interpretations of what they had agreed to. As Wood has highlighted, though, 'It is, of course, only possible to use clear language when the policy is clear' (1998: 82). The goal of 'doing something' can encourage states to paper over real divisions in policy using language that they can accept.

Internal legitimation practices—the drive towards unanimity, repetition of language, and regional support—and external legitimation practices—consistent arguments and 'doing something'—are all common and normalized within Security Council negotiations in the modern era. This book not only outlines evidence for these legitimation practices, but also demonstrates the impacts these had on specific negotiations on Darfur. The object of analysis, then, is not only the practices themselves but also how these informal and internal practices constrain and enable decision-making. The impact of legitimation practices does not necessarily lead to 'better' decisions in a normative sense or even decisions that are implementable. Indeed, legitimation practices can lead to perverse outcomes whereby seeking to make *a* decision can be prioritized over making a decision which can be implemented, as happened in relation to decisions on sanctions and UN peacekeeping in Darfur. These informal and often uncodified practices are critical, then, for understanding Security Council decision-making.

To put it into a familiar frame, these legitimation practices sit at the intersection of a logic of consequences and a logic of appropriateness (March and Olsen 1998, 2005). Although often taken as dichotomous positions, March and Olsen were clear from the outset that the two logics do not operate in isolation but that there is interaction between them. To take the drive towards unanimity as an example, it has a strategic component to it—unanimous resolutions are likely to be easier to enforce since they show greater political will—yet privileging unanimity leads to a sense of 'compliance pull' for states to vote affirmatively on resolutions, even when they have ongoing concerns. That is, privileging unanimity leads to a sense that the appropriate action for Security Council members is affirmative voting. As such, there are both strategic and appropriateness motivations in unanimous voting, which are difficult to separate. Similarly, seeking support from regionally relevant states could be motivated by a desire for the additional legitimacy, and perhaps enforcement benefits, which come from regional support; however, it could also be motivated by a sense that input from the region in question is normal and right. Arguments that seek legitimacy from specific audiences could be motivated by the strategic gains that could come from this (domestic support, buy-in from diplomatic partners etc.) or shifts in arguments could be motivated by new information and a change in perception on appropriate behaviour in the new situation. Either way, accessing motivations or intent is fraught; however, practices are public performances that we can access and assess.

Audiences warrant specific attention, given that audiences are intrinsic to the judgement of legitimacy. The Security Council is a discrete entity but one that is not isolated and interacts with other entities. Key audiences are fellow Security Council members where public and private comments are nominally directed, domestic audiences for whom arguments are sometimes relevant, especially in high-profile cases with strong domestic salience, and an international audience of peers. Even comments which are made in private need to reflect some awareness of public audiences, as there is always the possibility of private comments being revealed (Johnstone 2003; Xavier do Monte 2016). Evident in the case of Darfur are situations where arguments became untenable for democratic states with sensitivity to public opinion—such as the UK and the US opposing any discussion of Darfur because they were focused on the concurrent conflict between the north of Sudan and the southern region of Sudan (now South Sudan). As is shown in Chapter 4, the UK and US resisted any discussion of Darfur in the Security Council—despite possessing extensive intelligence of violence against civilians—but once it became a major national media story it was no longer tenable to focus solely on the north/south conflict. Fellow Security Council members, domestic publics, and international peers make up key audiences for Security Council activities.

Institutional power entrenches the privileges of permanent members, whose power goes beyond their formal Charter-derived privileges, as discussed in the previous chapter. Permanent members not only have the advantages of institutional memory, but also dominate drafting and agenda-setting, and have the capacity for informal vetoes. Their institutional power is not absolute, however, and is mitigated by both internal and external legitimation practices. External legitimation practices incentivize consistency in arguments and taking some form of action, because states seek the reputational benefits of maintaining or increasing status in the eyes of relevant domestic and international audiences. Internal legitimation practices incentivize a (somewhat) more collectivist approach to decision-making, since unanimity and regional support are valued. This means, for example, that although permanent members dominate drafting, internal legitimation practices still incentivize them, making the necessary adaptations to garner support from elected members and members from the region. As such, legitimation practices moderate the impacts of institutional power.

As the proceeding discussion shows, legitimation practices have a significant impact on how the Security Council conducts its negotiations on a day-to-day basis. While key national and material interests of permanent members play a role in setting 'red lines' for negotiations, much of what happens on a daily basis can be better understood via legitimation practices whereby states seek legitimacy for themselves as actors or the decision they are drafting. These patterns of behaviour are not static, and evolve over time, but are entrenched enough that they can be understood as practices that are meaningful to participants and their immediate epistemic community.

## Conclusion

Legitimation practices are a missing piece of the puzzle in understanding how the Security Council negotiates. While there has been a growing increase in interest in the relationship between the Security Council and legitimacy over the last two decades, the dominant approach has been to analyse the legitimacy of the institution or the legitimacy of its decisions. These traditional approaches consider how legitimacy functions outside the Security Council in terms of encouraging compliance with decisions, why powerful states might choose to channel policies through the Security Council, or whether the institution itself needs reform. By focusing inwards, this book instead demonstrates the importance of analysing how considerations of legitimacy shape the process of reaching a decision, and how this both enables and constrains which decisions are possible. Security Council members, and particularly the P3 which dominate drafting, prioritize both internal and external legitimation practices. None of these practices are inevitable, and they are also time specific. Indeed, during the Cold War, unanimity was neither achievable nor prioritized, and the importance of support from regional actors has grown in recent years. The formal rules set out in the UN Charter and the Provisional Rules of Procedure do not give any indication of these practices, which have developed informally over time. Although the Security Council has not codified these practices, members understand these practices as normal and repeat them across issue areas.

The bifurcated design of the UN Security Council, along with the institutional power that P5 members enjoy beyond their formal privileges, suggests a Council dominated by permanent members, as discussed in the previous chapter. Legitimation practices moderate the dominance of permanent members, however, as they incentivize broader engagement. If drafters want to achieve a unanimous vote, that gives them some incentive to engage with diverse perspectives. The drive towards 'doing something' can create situations where Security Council member make legally binding decisions which paper over real political divisions and, in doing so, pass decisions in the knowledge that they cannot be implemented. The focus of this book is not evaluating the decisions, however, but better understanding the decision-making process.

Part Two of the book analyses the Security Council's decision-making in response to the conflict in Darfur to demonstrate legitimation practices in action and to show how these practices shaped the process and outcome of negotiations. The structure is thematic, and largely chronological, with a chapter on each of the key tools that the Security Council applied to the conflict in Darfur. The next chapter outlines the situation on the ground in Darfur, as context for all following Security Council negotiations, and considers debates on both genocide and the responsibility to protect. The following chapters are thematic: agenda-setting, sanctions, referral to the International Criminal Court, and peacekeeping. A stronger understanding of how internal and external legitimation practices shape negotiations also opens up debates on informal reform of the Security Council's working methods, prospects for activism feeding into negotiations, and the relationship between national interests and actions within the Security Council.

# Bibliography

Adler, Emanuel and Vincent Pouliot. 2011a. 'International Practices'. *International Theory* 3(1): 1–36.

Adler, Emanuel and Vincent Pouliot. 2011b. 'International Practices: Introduction and Framework'. In *International Practices*, eds. Emanuel Adler and Vincent Pouliot. Cambridge: Cambridge University Press.

Adler-Nissen, Rebecca. 2015. 'Conclusion: Relationalism or Why Diplomats Find International Relations Theory Strange'. In *Diplomacy and the Making of World Politics*, eds. Ole Jacob Sending, Vincent Pouliot, and Ivan B. Neumann. Cambridge: Cambridge University Press.

Adler-Nissen, Rebecca and Alena Drieschova. 2019. 'Track-Change Diplomacy: Technology, Affordances, and the Practice of International Negotiations'. *International Studies Quarterly* 63(3): 531–545.

Adler-Nissen, Rebecca and Vincent Pouliot. 2014. 'Power in practice: Negotiating the international intervention in Libya'. *European Journal of International Relations* 20(4): 889–911.

Applbaum, Arthur Isak. 2000. 'Culture, Identity, and Legitimacy'. In *Governance in a Globalizing World*, eds. Joseph S. Nye and John D. Donahue: Brookings Institution Press.

Barker, Rodney. 2001. *Legitimating Identities: The Self-Presentation of Rules and Subjects*. Cambridge: Cambridge University Press.

Barnett, Michael N. 1997. 'The UN Security Council, Indifference, and Genocide in Rwanda'. *Cultural Anthropology* 12(4): 551–578.

Bellamy, Alex J. and Paul D. Williams. 2011. 'The New Politics of Protection? Cote d'Ivoire, Libya and the Responsibility to Protect'. *International Affairs* 87(4): 845–870.

Binder, Martin and Monika Heupel. 2014. 'The Legitimacy of the UN Security Council: Evidence from recent General Assembly debates'. *International Studies Quarterly* 59(2): 238–250.

Bode, Ingvild. 2018. 'Reflective Practices at the Security Council: Children and Armed Conflict and the Three United Nations'. *European Journal of International Relations* 24(2): 293–318.

Bode, Ingvild and John Karlsrud. 2019. 'Implementation in Practice: The Use of Force to Protect Civilians in United Nations Peacekeeping'. *European Journal of International Relations* 25(2): 458–485.

Booth Walling, Carrie. 2013. *All Necessary Measures: The United Nations and Humanitarian Intervention*. Pennsylvania Studies in Human Rights. Philadelphia: University of Pennsylvania Press.

Buchanan, Allen and Robert O. Keohane. 2006. 'The Legitimacy of Global Governance Institutions'. *Ethics and International Affairs* 20(4): 405–437.

Bueger, Christian. 2014. 'Pathways to Practice: Praxiography and International Politics'. *European Political Science Review* 6(3): 383–406.

Bueger, Christian and Frank Gadinger. 2018. *International Practice Theory*, 2 ed. Basingstoke and New York: Palgrave Macmillan.

Byers, Michael. 2004. 'Agreeing to Disagree: Security Council Resolution 1441 and Intentional Ambiguity'. *Global Governance* 10(2): 165–186.

Chapman, Terrence L. 2007. 'International Security Institutions, Domestic Politics, and Institutional Legitimacy'. *Journal of Conflict Resolution* 51(1): 134–166.

Chapman, Terrence L. 2009. 'Audience Beliefs and International Organizational Legitimacy'. *International Organization* 63(Fall): 733–764.

Chapman, Terrence L. and Dan Reiter. 2004. 'The United Nations Security Council and the Rally 'Round the Flag Effect'. *Journal of Conflict Resolution* 48(6): 886–909.

Chesterman, Simon. 2006. 'Reforming the United Nations: Legitimacy, Effectiveness and Power after Iraq'. *Singapore Year Book of International Law* 10: 1–28.

Clark, Ian. 2005. *Legitimacy in International Society*. New York: Oxford University Press.

Coleman, Katharina P. 2007. *International Organisations and Peace Enforcement: The Politics of International Legitimacy*. Cambridge: Cambridge University Press.

Dörfler, Thomas. 2019. *Security Council Sanctions Governance: The Power and Limits of Rules*. Routledge Research on the United Nations. Abingdon and New York: Routledge.

DPKO. 2008. *United Nations Peacekeeping Operations: Principles and Guidelines*. United Nations. Available at https://peacekeeping.un.org/sites/default/files/capstone_eng_0.pdf.

Dunne, Tim and Jess Gifkins. 2011. 'Libya and the State of Intervention'. *Australian Journal of International Affairs* 65(5): 515–529.

Elliott, Kimberly Ann. 2009–10. 'Assessing UN Sanctions After the Cold War'. *International Journal* 65(1): 85–97.

Elster, Jon. 1995. 'Strategic Uses of Argument'. In *Barriers to Conflict Resolution*, eds. Kenneth J. Arrow, Robert H. Mnookin, Lee Ross, Amos Tversky, and Robert B. Wilson. New York and London: W.W. Norton & Company.

Elster, Jon. 1998. 'Deliberation and Constitution Making'. In *Deliberative Democracy*, ed. Jon Elster. Cambridge: Cambridge University Press.

Finnemore, Martha. 2003. *The Purpose of Intervention: Changing Beliefs about the Use of Force*. Cornell Studies in Security Affairs. New York: Cornell University Press.

Franck, Thomas M. 1990. *The Power of Legitimacy among Nations*. Oxford: Oxford University Press.

Frost, Mervyn. 2013. 'Legitimacy and International Organizations: The Changing Ethical Context'. In *Legitimating International Organizations*, ed. Dominik Zaum. Oxford: Oxford University Press.

Gehring, Thomas and Thomas Dörfler. 2019. 'Constitutive Mechanisms of UN Security Council Practices: Precedent Pressure, Ratchet Effect, and Council Action Regarding Intrastate Conflicts'. *Review of International Studies* 45(1): 120–140.

Gifkins, Jess. 2016a. 'Naming and Framing: Darfur, the Genocide Debate, and the Responsibility to Protect'. In *The United Nations and Genocide*, ed. Deborah Mayerson. Basingstoke: Palgrave Macmillan.

Gifkins, Jess. 2016b. 'R2P in the UN Security Council: Darfur, Libya and Beyond'. *Cooperation and Conflict* 51(2): 148–165.

Gifkins, Jess, Samuel Jarvis and Jason Ralph. 2019. 'Brexit and the UN Security Council: Declining British Influence?'. *International Affairs* 95(6): 1349–1368.

Groom, A. J. R. and Sally Morphet. 2006. 'Baghdad to Baghdad: The United Kingdom's odyssey'. In *The Iraq Crisis and World Order: Structural, Institutional and Normative Challenges*, ed. Ramesh and Waheguru Pal Singh Sidhu Thakur. Tokyo: United Nations University Press.

Guéhenno, Jean-Marie. 2015. *The Fog of Peace: A Memoir of International Peacekeeping in the 21st Century*. Washington D.C.: The Brookings Institution.

Halliday, Terrence C., Susan Block-Lieb and Bruce G. Carruthers. 2010. 'Rhetorical Legitimation: Global Scripts as Strategic Devices of International Organizations'. *Socio-Economic Review* 8(1): 77–112.

Hopf, Ted. 2010. 'The logic of habit in International Relations'. *European Journal of International Relations* 16(4): 539–561.

Hopf, Ted. 2017. 'Change in International Practices'. *European Journal of International Relations* 24(3): 687–711

Howard, Lise Morjé and Anjali Kaushlesh Dayal. 2017. 'The Use of Force in UN Peacekeeping'. *International Organization* 72(1): 71–103.

Hulton, Susan C. 2004. 'Council Working Methods and Procedure'. In *The UN Security Council: From the Cold War to the 21st Century*, ed. David M. Malone. Boulder and London: Lynne Rienner Publishers, Inc.

Hurd, Ian. 1997. 'Security Council Reform: Informal Membership and Practice'. In *The Once and Future Security Council*, ed. Bruce Russett. New York: St. Martin's Press.

Hurd, Ian. 1999. 'Legitimacy and Authority in International Politics'. *International Organization* 53(2): 379–408.

Hurd, Ian. 2002. 'Legitimacy, Power, and the Symbolic Life of the UN Security Council'. *Global Governance* 8(1): 35–51.

Hurd, Ian. 2007. *After Anarchy: Legitimacy and Power in the United Nations Security Council*. Princeton: Princeton University Press.

Hurd, Ian. 2008. 'Myths of Membership: The Politics of UN Security Council Reform'. *Global Governance* 14: 199–217.

Hurrell, Andrew. 2005. 'Legitimacy and the Use of Force: Can the Circle Be Squared?'. *Review of International Studies* 31(1): 15–32.

Johnson, Cathryn, Timothy J. Dowd, and Cecilia L. Ridgeway. 2006. 'Legitimacy as a Social Process'. *Annual Review of Sociology* 32(1): 53–78.

Johnston, Alastair Iain. 2001. 'Treating International Institutions as Social Environments'. *International Studies Quarterly* 45: 487–515.

Johnstone, Ian. 2003. 'Security Council Deliberations: The Power of the Better Argument'. *European Journal of International Law* 14(3): 437–480.

Kratochwil, Friedrich. 2006. 'On Legitimacy'. *International Relations* 20(3): 302–308.

Krebs, Ronald R. and Patrick Thaddeus Jackson. 2007. 'Twisting Tongues and Twisting Arms: The Power of Political Rhetoric'. *European Journal of International Relations* 13(1): 35–66.

Lanz, David. 2020. *The Responsibility to Protect in Darfur: From Forgotten Conflict to Global Cause and Back*. Global Politics and the Responsibility to Protect. Abingdon and New York: Routledge.

Luck, Edward C. 2008. 'A Council for All Seasons: The Creation of the Security Council and Its Relevance Today'. In *The United Nations Security Council and War*, eds. Vaughan Lowe, Adam Roberts, Jennifer Welsh, and Dominik Zaum. Oxford: Oxford University Press.

Lynch, Colum. 2010. *The 10 Worst UN Security Council Resolutions Ever*. Turtle Bay, Foreign Policy. Available at http://turtlebay.foreignpolicy.com/posts/2010/05/21/the_10_worst_un_security_council_resolutions_ever.

March, James G. and Johan P. Olsen. 1998. 'The Institutional Dynamics of International Political Orders'. *International Organization* 52(4): 943–969.

March, James G. and Johan P. Olsen. 2005. 'The Logic of Appropriateness'. In *The Oxford Handbook of Public Policy*, eds. Michael Moran, Martin Rein, and Robert E. Goodin. Oxford: Oxford University Press.

Matsumura, Naoko and Atsushi Tago. 2019. 'Negative Surprise in UN Security Council Authorization: UK and French Vetoes Send Valuable Information for the General Public in Deciding If They Support a US Military Action'. *Journal of Peace Research* 56(3): 395–409.

Mills, Kurt. 2009. 'Vacillating on Darfur: Responsibility to Protect, to Prosecute, or to Feed?'. *Global Responsibility to Protect* 1(4): 532–559.

Morris, Justin. 2011. 'How Great is Britain? Power, Responsibility and Britain's Future Global Role'. *The British Journal of Politics & International Relations* 13(3): 326–347.

Morris, Justin and Nicholas J. Wheeler. 2007. 'The Security Council's Crisis of Legitimacy and the Use of Force'. *International Politics* 44: 214–231.

Natsios, Andrew. 2012. *Sudan, South Sudan, and Darfur: What Everyone Needs to Know*. Oxford: Oxford University Press.

Neumann, Ivan B. 2002. 'Returning Practice to the Linguistic Turn: The Case of Diplomacy'. *Millennium: Journal of International Studies* 31(3): 627–651.

Neumann, Ivan B. 2007. '"A Speech That the Entire Ministry May Stand For," Or: Why Diplomats Never Produce Anything New'. *International Political Sociology* 1(2): 183–200.

Niemann, Holger. 2022. 'Responsibility as Practice: Implications of UN Security Council Responsibilization'. In *The Routledge Handbook on Responsibility in International Relations*, eds. Hannes Hansen-Magnusson and Antje Vetterlein. Abingdon: Routledge.

Niemetz, Martin Daniel. 2015. *Reforming UN Decision-Making Procedures: Promoting a Deliberative System for Global Peace and Security*. Routledge Research on the United Nations. Abingdon and New York: Routledge.

OCHA. 2016. *Aide Memoire for the Consideration of Issues Pertaining to the Protection of Civilians in Armed Conflict*. Available at https://www.unocha.org/sites/unocha/files/Aide%20Memoire%202016%20II_0.pdf.

Oliver, Tim and Michael John Williams. 2016. 'Special Relationships in Flux: Brexit and the Future of the US-EU and US-UK Relationships'. *International Affairs* 92(3): 547–567.

Pouliot, Vincent. 2008. 'The Logic of Practicality: A Theory of Practice of Security Communities'. *International Organization* 62(02): 257–288.

Pouliot, Vincent. 2016. *International Pecking Orders: The Politics and Practices of Multilateral Diplomacy*. Cambridge: Cambridge University Press.

Pouliot, Vincent. 2020. 'The Gray Area of Institutional Change: How the Security Council Transforms Its Practices on the Fly'. *Journal of Global Security Studies* 6(3): 1–18.

Ralph, Jason and Adrian Gallagher. 2015. 'Legitimacy Faultlines in International Society: The Responsibility to Protect and Prosecute after Libya'. *Review of International Studies* 41(3): 553–573.

Ralph, Jason and Jess Gifkins. 2017. 'The Purpose of United Nations Security Council Practice: Contesting Competence Claims in the Normative Context Created by the Responsibility to Protect'. *European Journal of International Relations* 23(3): 630–653.

Ralph, Jason, Jess Gifkins and Samuel Jarvis. 2020. 'The UK's Special Responsibilities at the United Nations: Diplomatic Practice in Normative Context'. *The British Journal of Politics & International Relations* 22(2): 164–181.

Rosenthal, Gert. 2017. *Inside the United Nations: Multilateral Diplomacy up Close*. Abingdon and New York: Routledge.

Scharpf, Fritz. 1999. *Governing in Europe: Effective and Democratic?* Oxford: Oxford University Press.

Schia, Niels Nagelhus. 2013. 'Being Part of the Parade—"Going Native" in the United Nations Security Council'. *Political and Legal Anthropology Review* 36(1): 138–156.

Schia, Niels Nagelhus. 2017. 'Horseshoe and Catwalk: Power, Complexity, and Consensus-Making in the United Nations Security Council'. In *Palaces of Hope: The Anthropology of Global Organizations*, eds. Ronald Niezen and Marie Sapignoli. Cambridge: Cambridge University Press.

Schia, Niels Nagelhus and Ole Jacob Sending. 2015. 'Status and Sovereign Equality'. In *Small State Status Seeking: Norway's Quest for International Standing*, eds. Benjamin de Carvalho and Ivan B. Neumann. London and New York: Routledge.

Schimmelfennig, Frank. 2001. 'The Community Trap: Liberal Norms, Rhetorical Action, and the Eastern Enlargement of the European Union'. *International Organization* 55(1): 47–80.

Security Council Report. 2018a. *The Penholder System*. New York. Available at https://www.securitycouncilreport.org/atf/cf/%7B65BFCF9B-6D27-4E9C-8CD3-CF6E4FF96FF9%7D/Penholders.pdf.

Security Council Report. 2018b. *Security Council Working Methods: Provisional Progress*. Available at http://www.securitycouncilreport.org/atf/cf/%7B65BFCF9B-6D27-4E9C-8CD3-CF6E4FF96FF9%7D/research_report_working_methods_2018.pdf.

Security Council Report. 2020. *Lead Roles with the Council in 2020: Penholders and Chairs of Subsidiary Bodies*. Available at https://www.securitycouncilreport.org/monthly-forecast/2020-02/lead-roles-within-the-council-in-2020-penholders-and-chairs-of-subsidiary-bodies.php.

Security Council Report. 2022. *2022 Chairs of Subsidiary Bodies and Penholders*. Available at https://www.securitycouncilreport.org/atf/cf/%7B65BFCF9B-6D27-4E9C-8CD3-CF6E4FF96FF9%7D/working_methods_penholders_chairs.pdf.

Sievers, Loraine and Sam Daws. 2019. 'Unprecedented "Twinned" Council Presidencies of France and Germany Portend Risks and Rewards'. Update website of 'The Procedure of the UN Security Council'. Available at https://www.scprocedure.org/chapter-3-section-1f.

Smith, Courtney B. 2006. *Politics and Process at the United Nations: The Global Dance*. London: Lynn Rienner Publishers.

Spandler, Kilian. 2020. 'UNAMID and the Legitimation of Global-Regional Peacekeeping Cooperation: Partnership and Friction in UN-AU Relations'. *Journal of Intervention and Statebuilding* 14(2): 187–203.

Steffek, Jens. 2003. 'The Legitimation of International Governance: A Discourse Approach'. *European Journal of International Relations* 9(2): 249–275.

Tallberg, Jonas and Michael Zürn. 2019. 'The Legitimacy and Legitimation of International Organisations: Introduction and Framework'. *The Review of International Organizations* 14: 581–606.

Thompson, Alexander. 2006. 'Coercion Through IOs: The Security Council and the Logic of Information Transmission'. *International Organization* 60(1): 1–34.

Thompson, Alexander. 2010. *Channels of Power: The UN Security Council and US Statecraft in Iraq*. New York: Cornell University Press.

Tierney, Dominic. 2005. 'Irrelevant or Malevolent? UN Arms Embargoes in Civil Wars'. *Review of International Studies* 31(4): 645–664.

United Nations. 2004. *5015th Meeting United Nations Security Council*. New York, 30 July, S/PV.5015.

United Nations. 2013. *6943rd Meeting United Nations Security Council*. New York, S/PV.6943.

United Nations. 2022. UN Security Council Meetings & Outcomes Tables Dag Hammarskjöld Library. Available at https://research.un.org/en/docs/sc/quick/meetings/2022.

United Nations Security Council. 2017. *Note by the President of the Security Council*. Available at http://undocs.org/S/2017/507.

Visoka, Gëzim. 2020. 'International Intervention and Relational Legitimacy'. In *Local Legitimacy and International Peacebuilding*, eds. Oliver P. Richmond and Roger Mac Ginty. Edinburgh : Edinburgh University Press.

Voeten, Erik. 2001. 'Outside Options and the Logic of Security Council Action'. *American Political Science Review* 95(4): 845–858.

Voeten, Erik. 2005. 'The Political Origins of the UN Security Council's Ability to Legitimize the Use of Force'. *International Organization* 59(3): 527–557.

von Billerbeck, Sarah. 2019. '"Mirror, Mirror on the Wall:" Self-Legitimation by International Organizations'. *International Studies Quarterly* 64(1): 207–219.

von Billerbeck, Sarah. 2020. 'No Action Without Talk? UN Peacekeeping, Discourse, and Institutional Self-Legitimation'. *Review of International Studies* 46(4): 477–494.

von Einsiedel, Sebastian, David M. Malone, and Bruno Stagno Ugarte. 2015. *The UN Security Council in an Age of Great Power Rivalry*. United Nations University. Available at https://i.unu.edu/media/cpr.unu.edu/attachment/1569/WP04_UNSCAgeofPowerRivalry.pdf.

Wallensteen, Peter and Patrik Johansson. 2016. 'The UN Security Council: Decisions and Actions'. In *The UN Security Council in the 21st Century*, eds. Sebastian Von Einsiedel, David M. Malone, and Bruno Stagno Ugarte. Boulder and London: Lynne Rienner Publishers.

Washburne, Sarah. 2009. 'A Sudanese Identity Crisis?: A Discourse Analysis of the Bashir Regime's Legitimacy Problem, 1999–2008'. *Kharmasin Journal: Reflections on the Social and Political* Spring(03): 63–75.

Weiss, Thomas G. 2014. 'Military Humanitarianism: Syria Hasn't Killed It'. *The Washington Quarterly* 37(1): 7–20.

Welsh, Jennifer and Dominik Zaum. 2013. 'Legitimation and the UN Security Council'. In *Legitimating International Organizations*, ed. Dominik Zaum. Oxford: Oxford University Press.

Werner, Wouter. 2017. 'Recall It Again, Sam. Practices of Repetition in the Security Council'. *Nordic Journal of International Law* 86(2): 151–169.

Williams, Paul D. 2013. 'Regional and Global Legitimacy Dynamics: The United Nations and Regional Arrangements'. In *Legitimating International Organizations*, ed. Dominik Zaum. Oxford: Oxford University Press.

Wood, Michael C. 1998. 'The Interpretation of Security Council Resolutions'. In *Max Planck Yearbook of United Nations Law*, eds. Jochen A. Frowein and Rudiger Wolfrum. London: Kluwer Law International.

Wood, Michael C. 2016. 'The Interpretation of Security Council Resolutions, Revisited'. *The Max Planck Yearbook on the United Nations Law* 20: 3–35.

Xavier do Monte, Izadora. 2016. 'The Pen is Mightier than the H-Bomb'. *Interventions* 18(5): 669–686.

Yamashita, Hikaru. 2007. 'Reading "Threats to International Peace and Security" 1946–2005'. *Diplomacy and Statecraft* 18(3): 551–572.

Zaum, Dominik. 2013. 'International Organizations, Legitimacy, and Legitimation'. In *Legitimating International Organizations*, ed. Dominik Zaum. Oxford: Oxford University Press.

# PART TWO

PART TWO

# 3

# On the Ground in Darfur

*Given the structural turbulence of the Sudanese polity, perhaps the only surprise is that Darfur did not plunge into outright war earlier on.*

(de Waal 2007a: 35)

In the early 2000s, news outlets regularly described Darfur as the 'world's worst humanitarian crisis' (AFP 2004; BBC 2004). The conflict displaced millions of people, and hundreds of thousands of people either died as a direct result of the violence or through preventable illness exacerbated by widespread displacement. As a precursor to the conflict, the government in Khartoum had consolidated power and wealth in the north, and denied resources to the western region of Darfur. Attacks by rebel groups precipitated the conflict, whereupon the government responded with the same excessive violence that characterized its response to the concurrent conflict with the southern region of Sudan (now South Sudan). The conflict in Darfur generated considerable international interest, beyond that of comparable conflicts in Angola and the Democratic Republic of the Congo (Mamdani 2009b: 21). The development of R2P, which occurred around the same time, created significant interest in Darfur and raised expectations for international responses to mass atrocity crimes. Of particular focus in international debates on Darfur was the question of whether genocide was being committed, because there was an expectation that if governments labelled the situation genocide that it would mandate action (Brunk 2008; Gifkins 2016b; Glanville 2009). The intense civil society and foreign policy focus on Darfur meant that there were multiple audiences to be taken into consideration by states when setting out their positions in the Security Council (Elster 1995). International and domestic interest in the case of Darfur meant that there was a strong potential for external legitimation practices to shape negotiations, as demonstrated in subsequent chapters, while internal legitimation practices also shaped negotiations, as is normal in the Security Council.

This chapter provides background on what was happening in Darfur alongside the Security Council's negotiations and explains the intensity of international interest in this case by virtue of analogy with the Rwandan genocide and the parallel development of the responsibility to protect. To do so, the chapter proceeds in three sections. The first section charts the development of the conflict in Darfur

*Inside the UN Security Council.* Jess Gifkins, Oxford University Press. © Jess Gifkins (2023).
DOI: 10.1093/oso/9780192869029.003.0004

and its implications for the people of Darfur. The second section unpacks the question of genocide, which dominated much of the international debate on Darfur. The third section analyses how debates on R2P shaped national and international politics and meant that the conflict in Darfur gained high salience in Western countries. This unusually high level of international interest in the conflict created a situation where there was strong potential for external legitimation practices to be a feature of the Security Council's negotiations on Darfur.

## The Evolution and Development of the Conflict in Darfur

This section provides an overview of the conflict in Darfur as contextual background for the later analysis (for more background on the conflict, see in particular Cockett 2010; de Waal 2007b; Flint and de Waal 2008; Hassan and Ray 2009b; Natsios 2012; Prunier 2008). Darfur consists of the three western states of Sudan, covering an area about the size of France. Conflict in Darfur escalated in early 2003 after decades of extreme marginalization by the government in Khartoum, with lack of social services, healthcare, education, economic development, and infrastructure. In the late 1990s, poverty rates in Darfur were estimated at 97 per cent, with only 30 per cent of children enrolled in primary education and 1.6 doctors per 1000 people (Musa 2018: 26–27). Neighbouring foreign militaries from Chad and Libya used Darfur as a military base, meaning it was also a highly militarized society. Recent history in Darfur of drought from 1983 to 1984, and war between 1987 and 1989, and 1995 and 1999, had degraded the social fabric of Darfur and the capacity to deal with local disputes. The combination of marginalization, militarization, and unresolved conflicts over land and resources left the community in Darfur fragmented and tense. Indeed, the situation in Darfur was so dire that certain authors on Sudan have expressed surprise in hindsight that the situation didn't escalate into outright conflict sooner (de Waal 2007a: 35; Prunier 2008: 86).

There was low-level conflict in Darfur in 2001 and 2002, however, the key point of escalation was in early 2003.[1] Two rebel groups announced themselves in early 2003—the Justice and Equality Movement (JEM) and the Sudan Liberation Movement/Sudan Liberation Army (SLM/A). The two rebel groups together made clear gains on the government in April 2003 in a joint attack on al Fasher airport, where they destroyed planes and secured more vehicles and weapons. This attack changed the calculations of the Sudanese government, as it was the most damaging rebel attack it had experienced (Flint and de Waal 2008: 120–121). The government of Sudan and the Janjaweed militia responded to the uprising and early success of rebel groups with excessive military power and mass violence.

---

[1] Due to timing, this book does not cover the subsequent escalation of conflict in Sudan and Darfur in 2023.

The Darfur conflict is typically framed as an ethnic conflict between the Arab government in Khartoum and the Black African tribes in Darfur, which stands in contrast to the concurrent conflict between the Sudanese government and the south of Sudan (now South Sudan), which is framed as a religious conflict. The binary ethnic narrative of the conflict in Darfur was common both in Western media and in African media (Kothari 2010; Ray 2009). The ethnic narrative of conflict in Darfur is an oversimplification, however, as there were members of different ethnic groups on both sides of the conflict (Assal and I.D.F. 2010; Musa 2018). Ethnicity in Sudan is also understood as a more fluid concept that in the West; for example, it is possible to 'become Arab' by acquiring cattle (El-Tom 2009; Mahmoud 2004: 3). The framing of Black Africans in Darfur is a recent demarcation and one that historically did not exist (Lanz 2020: 123; Willemse 2009). The representation of this as an ethnic conflict also exacerbated the tensions between Arab and African groups in Sudan.

The rebel groups mainly comprised people from so-called Black African tribes, and the government responded by arming a pre-existing Arab militia known as the Janjaweed to suppress the rebellion. Alex de Waal aptly described the Sudanese government's actions in Darfur as 'counter-insurgency on the cheap', a strategy previously used to suppress rebel groups in the south of Sudan, which from the government's perspective gave it the added advantage of deniability and potential impunity (2004). The Janjaweed militia targeted civilian areas which were identified as 'Black' or 'non-Arab', with the aim of removing the support base for rebel groups by killing or displacing people of the same ethnic groups as the rebels. A pattern began whereby bombs were dropped from aircraft on civilian towns, then the Janjaweed would arrive on the ground, often with the regular army, where they would rape girls and women, loot, kill people, and burn down houses (Human Rights Watch 2004b: 15; Prunier 2008: 99–100).[2] The goal was not only to kill, but to destroy the requirements for life, often by putting corpses into wells and destroying food stores and tools needed for cooking (Flint and de Waal 2008: 145; Human Rights Watch 2004a: 41–42, 2004b: 33). Violence against civilians in Darfur, led by the Sudanese government and Janjaweed militia, decimated communities and previous sources of societal resilience.

The conflict escalated quickly, with dire consequences. As others have highlighted, the Security Council was slow to engage with the situation in Darfur (Abass 2007; Badescu and Bergholm 2009; Udombana 2005, 2007). The highest rates of violence and displacement were in the first year of the conflict. Mortality rates in Darfur have been highly contested and debated (for discussion see Mamdani 2009b: 26–34; Prunier 2008: 148–152). However, a study conducted

---

[2] There is minimal research available on the impact of the Darfur conflict on queer and gender non-conforming people, given the criminalization of sodomy and 'indecent acts' in Sudan, however, we know the conflict occurred alongside repression of LGBTQI+ people (Gifkins et al 2022). On queering responses to atrocity crimes see also (Gifkins and Cooper-Cunningham 2023).

by the United States Government Accountability Office (GAO) asked experts to determine their confidence levels in mortality figures from different sources, and analysts had the highest confidence in figures provided by the Centre for Research on the Epidemiology of Disasters (CRED) (GAO 2006).[3] The authors of the CRED report have since produced updated figures showing clearly that the height of violent deaths occurred from late 2003 through early 2004. Mortality rates in Darfur between 2003 and 2008 were high, with an estimated 300,000 'excess deaths' (Degomme and Guha-Sapir 2010: 297).[4] To understand what this tells us about the trajectory of the conflict, however, we need to separate the figures into violent deaths and deaths due to preventable illness. Of the 'excess deaths' the vast majority—80 per cent—were caused by diseases fuelled by displacement (primarily diarrhoea), and the remaining 20 per cent were a direct result of violence (Degomme and Guha-Sapir 2010: 297). These two different causes of death followed different patterns during the conflict. As shown in Figure 4, the highest rates of violent death occurred at the start of the conflict between early 2003 and early 2004. Significantly, this means that by the time the situation became part of the UN Security Council's agenda in May 2004, the highest levels of violence had passed.

Deaths due to illness were also high in the early stages of the conflict but took longer to stabilize. The international humanitarian response substantially increased through 2004 and 2005, and by 2005 overall mortality rates (predominately from illness) had dropped below what are considered to be 'emergency levels' and—although high—were comparable with mortality rates in Darfur prior

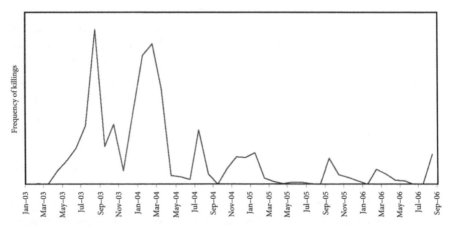

**Figure 4**  Frequency of Killings in Darfur over Time

*Source:* Adapted from ICC (2007)

---

[3] The analysis done by CRED was seen as the most credible, despite some weaknesses, because of its objectivity, careful explanations of its limitations, and innovative methods to triangulate data (for discussion see GAO 2006).

[4] 'Excess deaths' are those that occur earlier than the usual average life expectancy for a region.

to the war (Flint and de Waal 2008: 173; Garfield and Polonsky 2010: 501). The impact of the humanitarian response has been aptly described as the 'unsung success of the international response to Darfur' (Flint and de Waal 2008: 147). This drop in mortality was followed by another peak in deaths from preventable illness in the period July 2006 to September 2007, when decreased security led to both increased displacement and less humanitarian access (Degomme and Guha-Sapir 2010: 298). There was considerably less violence by this stage. Julie Flint, a key author on Darfur, reported that in July 2007, sixteen people were killed in the most significant violent incident for the month, not thousands as was the case during the height of the conflict, and that mortality levels were actually lower than before the war, despite advocacy through 2006 and 2007 calling for peacekeeping (2007). As Figure 5 shows, the first two years of the conflict caused about the same total level of displacement as the eight years from 2006 to 2013. The highest rates of displacement were in 2004 (OCHA 2009). Deaths due to violence decreased after the first phase of the conflict in 2003 and early 2004; however, displacement and insecurity led to ongoing deaths due to preventable diseases.

Critically, alongside the conflict in Darfur, there was a separate civil war between the northern government in Khartoum and the southern region of Sudan. This conflict spanned two decades and led to the deaths of two and a half million people (Natsios 2012: 1). An end to this civil war was negotiated during the early stages of the Darfur crisis and affected the Darfur conflict in three ways. First, the peace negotiations created expectations and costs for Darfuris in being left out of the negotiations (Hassan and Ray 2009a: 39; Seymour 2010: 59). Darfuri rebels realized that finalizing the north–south peace deal would exacerbate Darfur's access to representation and wealth and were resentful at being left out of these negotiations (Williams 2010). This increased the incentives for Darfuri rebels to

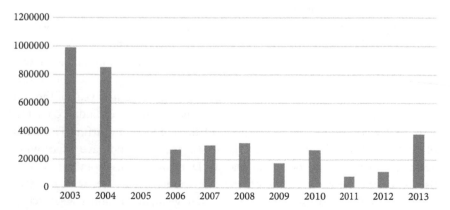

**Figure 5**  People Newly Displaced from Darfur

*Source:* Compiled from OCHA (2014)

*Note:* OCHA data did not contain figures for the number of people displaced in 2005.

try to gain ground militarily while the negotiations were taking place in the hope Darfur could be included in the peace negotiations. Second, the government of Sudan also had additional incentives to achieve its goals of suppressing the rebellion in Darfur before a peace deal was reached. After the peace deal was signed, the Khartoum government would form a joint government with the South, which they anticipated would make it harder to suppress the Darfuri rebellion (Williams 2010: 207). Third, and most crucially, while the Government of Sudan was negotiating with the international community towards a settlement for the South, it had free reign in Darfur (British International Development Committee 2005: 37; Cockett 2010: 241–242; Traub 2006). The international optimism on nearing a settlement between north and south meant that international participants were reluctant to pressure Khartoum on Darfur and risk jeopardizing the negotiations (British International Development Committee 2005: 35–36; Human Rights Watch 2004a: 50; ICG 2004a: 3, 2004b: ii; Kapila and Lewis 2013; Prunier 2008: 90). A sequential (and alternating) approach to the two conflicts in Sudan has been a key characteristic of the international response (Nitzschke 2016; Williams and Black 2010).

The US, the UK, and Norway were part of a troika that led the mediation of the north–south peace negotiations. They had considerable political capital invested in the process, so they were reluctant to jeopardize their relationship with the government of Sudan by criticizing its actions in Darfur. For the US government, ending the north–south civil war was a high foreign policy priority, connected to the Christian Right within the US and its advocacy on Sudan (Natsios 2012: 166–169). The Naivasha negotiations thus gave the Sudanese government and Darfuri rebels a closing window of opportunity in which to achieve their goals for Darfur and incentivized violence from both sides. In this way, the Naivasha negotiations both escalated the conflict in Darfur and limited the international community's willingness to engage with the crisis in Darfur.

The context of the conflict in Darfur meant that it was challenging for Security Council members from the start. Violence was government sponsored, meaning that Khartoum was unwilling to curtail the Janjaweed militia, and actively obstructed access to Darfur for media, NGOs, and UN agencies. The conflict escalated rapidly, with the worst violence occurring in the first year of the conflict. Peace negotiations between Khartoum and the south of Sudan coincided with, and exacerbated, the early stages of the conflict in Darfur. The parallel civil war diverted international diplomatic attention and gave Khartoum a free reign to commit atrocities in Darfur. Drawing lessons from past failures also had some perverse effects on the international response, with the erroneous assumption that 'naming' the situation 'genocide' would trigger action. These features of the situation on the ground in Darfur and the international context made it particularly difficult to garner international consensus and action.

## Darfur and the Question of Genocide

Media and advocacy groups regularly framed the conflict in Darfur as analogous to the Rwandan genocide, which had some perverse effects on the international response (Gifkins 2016b; Lanz 2011; Mamdani 2009b). Failure on the part of the international community to take action during the Rwandan genocide became linked to the failure to name the situation genocide while it was occurring (Brunk 2008; Glanville 2009). In March 2004, prominent *New York Times* journalist Nicolas Kristof called the situation in Darfur genocide (2004). This sparked a debate over whether the situation legally constituted genocide or not, which became a dominant question in the international response to Darfur, at times eclipsing calls for humanitarian aid, at a time when disease was the most common cause of death.

In July 2004, US Secretary of State Colin Powell said, 'we concluded, I concluded, that genocide has been committed in Darfur, and that the Government of Sudan and the Janjaweed bear responsibility and that genocide may still be occurring' (Committee on Foreign Relations 2004). This was the first time that the US Administration had described a situation as genocide while it was ongoing, which initially suggested that the US government might take a stronger approach to the crisis. However, Powell continued: 'no new action is dictated by this determination' (Committee on Foreign Relations 2004). A subsequent investigation by the International Commission of Inquiry on Darfur[5] determined that the requisite 'intent to destroy', which is a legal component of genocide, had not been found in Darfur, but it determined that war crimes and crimes against humanity had been committed, which it said 'may be no less serious and heinous than genocide' (United Nations 2005a).

Following these two pronouncements on the question of genocide, there was an investigation by the International Criminal Court.[6] In 2009, it issued an arrest warrant for the then Sudanese President, Oman Hassan Ahmad Al-Bashir, on charges of crimes against humanity and war crimes, and in 2010 a further arrest warrant was issued on the charge of genocide (Lynch and Hamilton 2010). These investigations, and the variation between them, maintained the prominence of the 'genocide question' as a key part of the international debates on Darfur. The multitude of investigations attempting to 'prove' genocide had occurred detracted resources from the humanitarian effort. The debates on genocide led some activists and humanitarian agencies to prioritize investigating genocide in 2005 and 2006, at a time when preventable illness was the biggest killer, over digging latrines and providing much-needed humanitarian assistance (Flint and de Waal 2008: 186; Weissman 2007). While activists thought that the label of genocide would garner

[5] The Commission of Inquiry on Darfur was established under Security Council resolution 1564, which is discussed in Chapter 5.

[6] The International Criminal Court's investigation of Darfur was mandated under Security Council resolution 1593, which is discussed in Chapter 6.

action, the focus on investigations to prove genocide—despite consistent and compelling evidence of crimes against humanity and war crimes—was often to the detriment of humanitarian action.

## Darfur and the Responsibility to Protect

The 2003 conflict in Darfur escalated during the period where the responsibility to protect, or 'R2P' as it is known, was in its initial stages of development. The International Commission on Intervention and State Sovereignty (ICISS) developed the idea of R2P in 2001, drawing on earlier ideas on 'sovereignty as responsibility' put forward by Francis Deng, a Sudanese diplomat and scholar, and his collaborators (Deng et al. 1996; ICISS 2001). African states and leaders played key roles in the development of R2P, most notably using language similar to R2P in the African Union's Constitutive Act in 2000 (African Union 2000; Hindawi 2022). The principle of R2P made remarkably quick progress from an idea in a report in 2001 to a principle agreed by all UN member states in 2005, as part of the World Summit outcome document, where R2P was agreed to apply to the prevention of genocide, war crimes, crimes against humanity, and ethnic cleansing (Bellamy 2009; Evans 2008). The 2005 version of R2P has three pillars. First, each governments' primary responsibility to protect their own people from the four atrocity crimes, and their incitement. Second, the role of the international community in assisting states to meet their obligations. Third, a role for the international community to take action via the UN Security Council if a state is manifestly failing to protect its people (Ban 2008, 2009). The key caveat for the Security Council, however, is that no specific actions are mandated, only that it considers issues on a 'case-by-case' basis, in line with the UN Charter, as it usually would (United Nations 2005b). As such, Jennifer Welsh, the former Special Advisor to the UN Secretary-General on R2P, has argued that this is best understood as a 'duty to consider' rather than mandating any particular outcome from Security Council negotiations (2013: 368). The extensive debates over R2P shaped debates over how the international community should respond to the crisis in Darfur.

The escalation of the conflict in Darfur occurred in parallel with the development and formalization of R2P. As I have argued elsewhere, there are some limitations in framing Darfur as a 'test case' for R2P, as was often done at the time, because by the time the R2P agreement was formalized in 2005, the height of the violence in Darfur had already receded (Gifkins 2016a). Despite not serving as a meaningful 'test case' of R2P, there is a clear relationship between R2P and the conflict in Darfur due to the parallel developments. After the ICISS report but before the formal agreement on R2P, there were some early instances invoking language from the ICISS report within Security Council debates and resolutions on Darfur (Williams and Bellamy 2005). Following the 2005 agreement on R2P, there were intense negotiations in the Security Council to reference the World Summit language on R2P in a resolution on the Protection of Civilians, championed by the UK, France, and Denmark, which was achieved in resolution 1674 (Gifkins 2016c: 153–155). The first country-specific use of R2P in a Security Council resolution was in resolution 1706 on Darfur, which was highly contentious and took six months of negotiation, championed again by the UK and France (Gifkins 2016c: 155–156).[7] Beyond the Security Council, R2P was instrumental in many aspects of the international politics and responses to the crisis in Darfur, from international calls for peacekeeping to local dynamics where the appropriation of R2P encouraged rebel groups to take maximalist positions (Lanz 2020). The parallel evolution of R2P alongside the Darfur conflict meant advocates and news outlets often viewed actions on Darfur through the prism of debates on R2P.

Connected to both R2P and the genocide debate, the conflict in Darfur generated significant civil society movements especially in the US and the UK. The Save Darfur movement in the US has been described as 'arguably the largest international social movement since anti-apartheid' (Lanz 2009: 669). The conflict in Darfur developed an unusual domestic salience in Western countries. For example, former US Deputy Secretary of State Robert Zoellick said, 'I have dealt with a lot of foreign policy issues ... and I am not sure I have one that has generated as much interest across a wide spectrum of Americans as this one has' (cited in Lanz 2011: 240). While this Western interest has been critiqued for oversimplifying and polarizing the conflict, it has also arguably generated traction internationally in a way that would likely not have happened otherwise (Lanz 2009; Mamdani 2009a). Western advocacy was often disconnected from the realities on the ground in Darfur, but it did generate attention within the Security Council, particularly on peacekeeping and referral to the International Criminal Court, that would likely not have occurred on that scale without such advocacy (Hamilton 2011). There was an unusually high level of Western interest and advocacy over the conflict in Darfur.

The eventual intensity of international interest in the conflict in Darfur meant that there was a significant audience interested in the Security Council's actions and considerable scope for external legitimation practices to shape the negotiations, as subsequent chapters will demonstrate. While internal legitimation practices are common across negotiations, external legitimation practices have a more direct relationship with external audiences which means that they can vary in intensity based on the external interest in a given case. Domestic interest in this case, especially in the US, was remarkably high, including, for example, a 'million voices' campaign, which was an initiative of the Save Darfur Coalition, to send postcards directly to President Bush calling for UN peacekeeping in Darfur (Hamilton 2011: 80–81). Another US NGO, Genocide Intervention Network,

---

[7] The negotiations towards resolution 1706 are discussed in more depth in Chapter 7.

created scorecards for members of congress from A+ to F based on their position on Darfur, which constituents then used in advocacy.[8] The scorecards incentivized members of Congress to be active on Darfur; constituents criticized those with low scores, as Hamilton explains. 'In some cases citizens used the scorecards in face-to-face meetings with representatives, whiles other published articles in their local papers to publicly shame low-scoring representatives from the area. Meanwhile, members of congress with high scores found themselves receiving calls of support from their constituents, calls that for the first time built a positive political incentive to respond to foreign atrocities' (2011: 100). As such, the external audience interest in Security Council actions on Darfur was unusually high, which meant that Security Council negotiators, particularly those from Western states, were more mindful than usual of how the Council's actions would look to domestic publics.

## Conclusion

As one of the worst humanitarian crises of its time, the conflict in Darfur gained significant domestic salience in Western countries, prompted by sustained media interest and through analogy to the Rwandan genocide. The parallel development of R2P energized many people in Western countries into a significant civil society movement aimed at protecting civilians in Darfur through advocacy and diplomacy. While many negotiations in the Security Council occur with limited international interest, negotiations on Darfur had an unusually high level of domestic interest, with advocacy groups calling for specific actions like UN peacekeeping. This meant that states, especially the Western members of the Security Council, were particularly mindful of how the domestic public and civil society organizations at home would perceive Security Council decisions. While external legitimation practices are common across Security Council negotiations, there was more scope for external legitimation practices than usual in a case like Darfur with particularly high domestic salience.

The conflict in Darfur had considerable media interest in Western countries, with grassroots advocacy organizations attempting to provoke multilateral action. In the US, constituents praised members of Congress for taking strong positions on Darfur and criticized those who did not engage. Despite the limitations in what these endeavours specifically achieved in terms of Security Council action, it was a remarkably coordinated attempt for Americans who may not have otherwise been active in politics to lobby for the Security Council to take specific actions. A meta study of key news outlets from around the world found that the most

---

[8] The Save Darfur Coalition and the Genocide Intervention Network merged in 2010 as a new organization called United to End Genocide.

comprehensive coverage came from the *Washington Post* and the British Broadcasting Corporation (BBC), with the *The New York Times* and the BBC providing the coverage with the highest emotional intensity (Mody 2010: 327 and 340). As one of the key media and political issues of its day, the conflict in Darfur had unusually high resonance in Western states, which meant that there was awareness that UN activity would face higher levels of domestic scrutiny than usual. The conflict in Darfur and the intense media and public interest it received meant that there was fertile ground for external legitimation practices in the Security Council to shape negotiations.

# Bibliography

Abass, Ademola 2007. 'The United Nations, the African Union and the Darfur Crisis: of Apology and Utopia'. *Netherlands International Law Review* 54: 415–440.

AFP. 2004. *UN Human Rights Forum to Turn Spotlight on Alleged Atrocities in Sudan.* Reliefweb. Available at http://reliefweb.int/node/145981.

African Union. 2000. *Constitutive Act of the African Union.* Togo, 11 July. Available at http://www.au2002.gov.za/docs/key_oau/au_act.htm.

Assal, Munzoul and I.D.F. 2010. 'The National Congress Party and the Darfurian Armed Groups'. In *The International Politics of Mass Atrocities: The Case of Darfur,* eds. David R. Black and Paul D. Williams. London and New York: Routledge.

Badescu, Cristina G. and Linnéa Bergholm. 2009. 'The Responsibility to Protect and the Conflict in Darfur: The Big Let-Down'. *Security Dialogue* 40(3): 287–309.

Ban, Ki-moon. 2008. *Secretary-General Defends, Clarifies 'Responsibility to Protect' at Berlin Event on 'Responsible Sovereignty: International Cooperation for a Changed World'.* Berlin. Available at http://www.un.org/News/Press/docs/2008/sgsm11701. doc.htm.

Ban, Ki-moon. 2009. *Implementing the responsibility to protect: report of the Secretary-General.* United Nations General Assembly, 12 January, A/63/677.

BBC. 2004. *Mass Rape Atrocity in West Sudan.* Available at http://news.bbc.co.uk/2/ hi/africa/3549325.stm.

Bellamy, Alex J. 2009. *Responsibility to Protect: The Global Effort to End Mass Atrocities.* Cambridge: Polity Press.

British International Development Committee. 2005. *Darfur, Sudan: The Responsibility to Protect.* House of Commons. Available at http://www.publications.parliament. uk/pa/cm200405/cmselect/cmintdev/67/67i.pdf.

Brunk, Darren. 2008. 'Disecting Darfur: Anatomy of a Genocide Debate'. *International Relations* 22(1): 25–44.

Cockett, Richard. 2010. *Sudan: Darfur and the Failing of an African State.* New Haven and London: Yale University Press.

Committee on Foreign Relations. 2004. *The Current Situation in Sudan and the Prospects for Peace.* United States Senate, 108th Congress, 9 September. Available at http://www.gpo.gov/fdsys/pkg/CHRG-108shrg97822/html/CHRG-108shrg97822.htm.

de Waal, Alex. 2004. *Counter-Insurgency on the Cheap.* Available at http://www.lrb.co. uk/v26/n15/alex-de-waal/counter-insurgency-on-the-cheap.

de Waal, Alex. 2007a. 'Sudan: The Turbulent State'. In *War in Darfur and the Search for Peace*, ed. Alex de Waal. Global Equity Initiative, Harvard University and Justice Africa.

de Waal, Alex, ed. 2007b. *War in Darfur and the Search for Peace*. Global Equity Initiative, Harvard University and Justice Africa.

Degomme, Olivier and Debarati Guha-Sapir. 2010. 'Patterns of Mortality Rates in Darfur Conflict'. *Lancet* 375: 294–300.

Deng, Francis M., Sadikiel Kimaro, Terrence Lyons, Donald Rothchild, and I. William Zartman. 1996. *Sovereignty as Responsibility: Conflict Management in Africa*. Brookings Institution Press.

El-Tom, Abdullahi Osman. 2009. 'Darfur People: Too Black for the Arab-Islamic Project of Sudan'. In *Darfur and the Crisis of Governance in Sudan: A Critical Reader*, ed. Salah M. Hassan and Carina E. Ray. Ithaca and London: Cornell University Press.

Elster, Jon. 1995. 'Strategic Uses of Argument'. In *Barriers to Conflict Resolution*, eds. Kenneth J. Arrow, Robert H. Mnookin, Lee Ross, Amos Tversky, and Robert B. Wilson. New York and London: W.W. Norton & Company.

Evans, Gareth. 2008. *The Responsibility to Protect: Ending Mass Atrocity Crimes Once and For All*. Washington D.C.: Brookings Institution Press.

Flint, Julie. 2007. 'Darfur: All this Moral Posturing Won't Help'. *The Independent*. 31 july.

Flint, Julie and Alex de Waal. 2008. *Darfur: A New History of a Long War*. African Arguments. London and New York: Zed Books.

GAO. 2006. *Darfur Crisis: Death Estimates Demonstrate Seversity of Crisis, but Their Accuracy and Credibility Could Be Enhanced*. Available at http://www.gao.gov/new.items/d0724.pdf.

Garfield, Richard and Jonny Polonsky. 2010. 'Changes in Mortality Rates and Humanitarian Conditions in Darfur, Sudan 2003–2007'. *Prehospital and Disaster Medicine* 25(6): 496–502.

Gifkins, Jess. 2016a. 'Darfur'. In *Oxford Handbook of the Responsibility to Protect*, eds. Alex J. Bellamy and Tim Dunne. Oxford: Oxford University Press.

Gifkins, Jess. 2016b. 'Naming and Framing: Darfur, the genocide debate, and the Responsibility to Protect'. In *The United Nations and Genocide*, ed. Deborah Mayerson. Basingstokes: Palgrave Macmillan.

Gifkins, Jess. 2016c. 'R2P in the UN Security Council: Darfur, Libya and Beyond'. *Cooperation and Conflict* 51(2): 148–165.

Gifkins, Jess, Dean Cooper-Cunningham, Kate Ferguson, Detmer Kremer and Farida Mostafa. 2022. Queering Atrocity Prevention. Available at https://protectionapproaches.org/queeringap.

Gifkins, Jess and Dean Cooper-Cunningham. 2023. 'Queering the Responsibility to Protect'. *International Affairs* OnlineFirst.

Glanville, Luke. 2009. 'Is "Genocide" Still a Powerful Word?'. *Journal of Genocide Research* 11(4): 467–486.

Hamilton, Rebecca. 2011. *Fighting for Darfur: Public Action and the Struggle to Stop Genocide*. New York: Palgrave Macmillan.

Hassan, M. Salah and Carina E. Ray. 2009a. 'Introduction: Critically Reading Darfur and the Crisis of Governance in Sudan'. In *Darfur and the Crisis of Governance in Sudan: A Critical Reader*, ed. Salah M. Hassan and Carina E. Ray. Ithaca and London: Cornell University Press.

Hassan, M. Salah and Carina E. Ray, eds. 2009b. *Darfur and the Crisis of Governance in Sudan*. New York: Cornell University Press.

Hindawi, Coralie Pison. 2022. 'Decolonizing the Responsibility to Protect: On pervasive Eurocentrism, Southern agency and struggles over universals'. *Security Dialogue* 53(1): 38–56.

Human Rights Watch. 2004a. *Darfur Destroyed: Ethnic Cleansing by Government and Militia Forces in Western Sudan*. Available at http://www.hrw.org/sites/default/files/reports/sudan0504full.pdf.

Human Rights Watch. 2004b. *Sudan, Darfur in Flames: Atrocities in Western Sudan*. Available at http://www.hrw.org/sites/default/files/reports/sudan0404.pdf.

ICC. 2007. *Prosecutor's Application under Article 58(7), Annex 3*. Available at https://www.icc-cpi.int/sites/default/files/RelatedRecords/CR2007_02091.PDF.

ICG. 2004a. *Darfur Rising: Sudan's New Crisis*. International Crisis Group. Available at http://www.crisisgroup.org/en/regions/africa/horn-of-africa/sudan/076-darfur-rising-sudans-new-crisis.aspx.

ICG. 2004b. *Sudan: Now or Never in Darfur*. International Crisis Group.

ICISS. 2001. *The Responsibility to Protect: The Report of the International Commission on Intervention and State Sovereignty*. Ottawa: International Development Research Centre.

Kapila, Mukesh and Damien Lewis. 2013. *Against a Tide of Evil: How One Man Became the Whistleblower to the First Mass Murder of the Twenty-First Century*. Edinburgh and London: Mainstream Publishing.

Kothari, Ammina. 2010. 'The Framing of the Darfur Conflict in the New York Times: 2003–2006'. *Journalism Studies* 11(2): 209–224.

Kristof, Nicholas D. 2004. 'Will We Say "Never Again" Yet Again?'. *The New York Times*. 27 March.

Lanz, David. 2009. 'Commentary—Save Darfur: A Movement and Its Discontents'. *African Affairs* 108(433): 669–677.

Lanz, David. 2011. 'Why Darfur? The Responsibility to Protect as a Rallying Cry for Transnational Advocacy Groups'. *Global Responsibility to Protect* 3(2): 223–247.

Lanz, David. 2020. *The Responsibility to Protect in Darfur: From Forgotten Conflict to Global Cause and Back*. Global Politics and the Responsibility to Protect. Abingdon and New York: Routledge.

Lynch, Colum and Rebecca Hamilton. 2010. *International Criminal Court Charges Sudan's Omar Hassan al-Bashir with Genocide*. Available at http://www.washingtonpost.com/wp-dyn/content/article/2010/07/12/AR2010071205295.html.

Mahmoud, Mahgoub El-Tigani. 2004. 'Inside Darfur: Ethnic Genocide by a Governance Crisis'. *Comparative Studies of South Asia, Africa and the Middle East* 24(2): 3–17.

Mamdani, Mahmood. 2009a. 'The Politics of Naming: Genocide, Civil War, Insurgency'. In *Darfur and the Crisis of Governance in Sudan: A Critical Reader*, ed. Salah M. Hassan and Carina E. Ray. Ithaca and London: Cornell University Press.

Mamdani, Mahmood. 2009b. *Saviors and Survivors: Darfur, Politics and the War on Terror*. New York: Pantheon Books.

Mody, Bella. 2010. *The Geopolitics of Representation in Foreign News: Explaining Darfur*. Plymouth: Lexington Books.

Musa, Suad M.E. 2018. *Hawkes and Doves in Sudan's Armed Conflict: Al-Hakkamat Baggara Women of Darfur.* Eastern Africa Series. Rochester: Boydell & Brewer Publishers.

Natsios, Andrew. 2012. *Sudan, South Sudan, and Darfur: What Everyone Needs to Know.* Oxford: Oxford University Press.

Nitzschke, Heiko. 2016. 'Sudan'. In *The UN Security Council in the 21st Century,* eds. Sebastian Von Einsiedel, David M. Malone, and Bruno Stagno Ugarte. Boulder and London: Lynne Rienner Publishers.

OCHA. 2009. *Darfur Humanitarian Profile No. 34.* Office of UN Deputy Special Representative of the UN Secretary-General for Sudan and UN Resident and Humanitarian Co-ordinator Available at http://reliefweb.int/sites/reliefweb.int/files/resources/D85EAF37950F12548525758B006DF908-Full_Report.pdf.

OCHA. 2014. *Sudan: 2014 Population Displacement in Darfur.* Available at http://reliefweb.int/sites/reliefweb.int/files/resources/OCHA%20Darfur%20Update%20-%2026%20May%202014.pdf.

Prunier, Gerard. 2008. *Darfur: A 21st Century Genocide,* 3rd ed. Crises in World Politics. New York: Cornell University Press.

Ray, Carina E. 2009. 'Darfur in the African Press'. In *Darfur and the Crisis of Governance in Sudan: A Critical Reader,* ed. Salah M. Hassan and Carina E. Ray. Ithaca and London: Cornell University Press.

Seymour, Lee J.M. 2010. 'The Regional Politics of the Darfur Crisis'. In *The International Politics of Mass Atrocities: The Case of Darfur,* eds. Paul D. Williams and David R. Black. London and New York: Routledge.

Traub, James. 2006. *The Best Intentions: Kofi Annan and the UN in the Era of American World Power.* New York: Farrar, Straus and Giroux.

Udombana, Nsongurua J. 2005. 'When Neutrality Is a Sin: The Darfur Crisis and the Crisis of Humanitarian Intervention in Sudan'. *Human Rights Quarterly* 27: 1149–1199.

Udombana, Nsongurua J. 2007. 'Still Playing Dice with Lives: Darfur and Security Council Resolution 1706'. *Third World Quarterly* 28(1): 97–116.

United Nations. 2005a. *Report of the International Commission of Inquiry on Darfur to the United Nations Secretary-General.* Geneva, 25 January. Available at http://www.un.org/news/dh/sudan/com_inq_darfur.pdf.

United Nations. 2005b. *World Summit Outcome—United Nations General Assembly.* A/RES/60/1.

Weissman, Fabrice. 2007. *Sudan Divided: The Challenge to Humanitarian Action.* New York: Medicins Sans Frontieres. Available at http://www.doctorswithoutborders.org/news-stories/speechopen-letter/sudan-divided-challenge-humanitarian-action.

Welsh, Jennifer. 2013. 'Norm Contestation and the Responsibility to Protect'. *Global Responsibility to Protect* 5(4): 365–396.

Willemse, Karin. 2009. 'The Darfur War: Masculinity and the Construction of a Sudanese National Identity'. In *Darfur and the Crisis of Governance in Sudan: A Critical Reader,* eds. Salah M. Hassan and Carina E. Ray. Ithaca and London: Cornell University Press.

Williams, Paul D. 2010. 'The United Kingdom'. In *The International Politics of Mass Atrocities: The Case of Darfur,* eds. Paul D. Williams and David R. Black. London and New York: Routledge.

Williams, Paul D. and Alex J. Bellamy. 2005. 'The Responsibility to Protect and the Crisis in Darfur'. *Security Dialogue* 36(1): 27–47.

Williams, Paul D. and David R. Black. 2010. 'Introduction: International Society and the Crisis in Darfur'. In *The International Politics of Mass Atrocities: The Case of Darfur* eds. Paul D. Williams and David R. Black. London and New York: Routledge.

# 4

# The Politics of Agenda-Setting

*Why is this focus on the weak little U.N. Secretariat, underfunded,
under-resourced, undermanned and without technical means that the
world's great powers have, actually to have real-time knowledge about
what's happening on the ground?*

Kieran Prendergast, former Under-Secretary-General
of the UN Department of Political Affairs, speaking
about Darfur (Frontline 2007a)

The conflict in Darfur escalated in 2003, however, the UN Security Council did
not engage with the crisis for the first year. It took more than twelve months for the
Security Council to include the situation in Darfur on its agenda, which is 'quick'
compared to conflicts in Uganda and South Sudan which took over a decade and
over two decades respectively, however, it was very slow for people on the ground
in Darfur (Binder and Golub 2020: 419). The highest intensity of conflict in Darfur passed before the Security Council took any actions to decrease violence, or
even made any statements about the crisis. This chapter analyses the year-long
process between the escalation of conflict in Darfur and the point where the Security Council added the situation in Darfur to its formal agenda. As highlighted
in the Introduction, analysis on the Security Council's engagement with Darfur
is often framed as Western advocates versus China. The reality is much more
nuanced than this; Western states actively blocked any discussion of Darfur in the
early stages of the conflict, the US was obstructionist over the later ICC referral,
and China constructively encouraged Khartoum to allow UN peacekeepers into
Darfur. This chapter demonstrates how it was external legitimation practices—
patterns of action that serve to legitimate an actor—that led Western states to
champion Darfur, rather than simply awareness about the scale and nature of the
violence.

Critical to the agenda-setting phase was US politics and the timing of public
awareness of the crisis. The US planned to normalize relations with Sudan after
parties to the conflict signed a peace agreement on the concurrent civil war in
Sudan between the northern government and the southern region of Sudan (now
South Sudan). US President Bush planned to fly key Sudanese officials to Washington DC to attend his State of the Union address as a celebration of the US' role

*Inside the UN Security Council.* Jess Gifkins, Oxford University Press. © Jess Gifkins (2023).
DOI: 10.1093/oso/9780192869029.003.0005

in the Sudanese peace deal. However, a statement to the media from a member of the UN Secretariat linked Darfur with the tenth anniversary of the Rwandan genocide triggered an exponential increase in media interest in Darfur and domestic awareness. This made it difficult for the US to continue to sideline Darfur while celebrating a peace deal in Sudan. It was this domestic interest, not simply the scale and nature of violence in Darfur, which triggered a change in position for the US, and similarly the UK. These states had extensive intelligence on the scale and nature of the violence in the early stages of the conflict and resisted action (Kapila and Lewis 2013). Following considerable media interest, however, the arguments they could make which would be understood as legitimate by key audiences changed. As such, it was external legitimation practices—states seeking legitimacy for their arguments, and by extension for themselves as actors—which enabled the situation in Darfur to become an item on the Security Council's agenda. This aligns with arguments made by Frederking and Patane that legitimacy-based accounts can be a strong indicator of which cases become part of the UN Security Council's agenda (2017).

The chapter begins with a background section that covers how the Security Council's agenda works. Unlike the later chapters that focus on interactions between states inside the Security Council, this chapter analyses the process by which Darfur became part of the Council's decision-making, so is necessarily more 'external' in orientation than subsequent chapters. The second section analyses Security Council inaction during the first six months of the conflict, when the barrier to action was availability of information, as the Sudanese government resisted access to media, UN agencies, and NGOs. The third section charts resistance from Western states on engaging with Darfur in the Security Council once information became available. The final section analyses the shift whereby media interest in Darfur increased and calculations of Western states changed. To be clear, the US and UK had detailed intelligence on the scale and nature of the violence against civilians in Darfur; their calculations changed once this information became publicly available. To enhance their legitimacy as actors, and to maintain a consistent argument as champions for Sudan, the UK and the US became advocates for Darfur.

## Agenda-Setting in the UN Security Council

Adding an item to the agenda is contentious both for the tools it enables and for the symbolic politics of legitimizing an issue as an appropriate matter for consideration by the Security Council. Even though permanent members have no enhanced formal role in agenda-setting compared to elected members, they still dominate as a function of their institutional power, whereby the relationships between permanent and elected members are mediated by their formal institutional differences.

Elected members can have more of a role in setting the agenda via the rotating presidency of the Security Council, whereby the president has can add items to the daily agenda (Allen and Yuen 2022). Adding an item to the agenda is generally slow, however, taking an average of more than three years between escalation of conflict and the Security Council adding it to their agenda (Binder and Golub 2020: 428). Boulden outlines three different scenarios that determine whether a situation becomes part of the Security Council's agenda. First, when one or more of the P5 have specific interests at stake, they either resist adding it to the agenda or seek to add it to achieve specific goals. Second, when the P5 have fewer interests but media or NGO advocacy motivates them to take action. Third, when the P5 do not have any interests, it does not become an agenda item (2006: 413–414). What Boulden's framework highlights is that, in the absence of strong interests from P5 members, media and NGO advocacy can determine whether or not the Security Council engages, as happened in the case of Darfur.

Permanent members and, more specifically, P3 members, dominate setting the Security Council's agenda as a function of their institutional power. Formally, permanent members hold no specific powers in relation to the agenda; the Secretary-General drafts the agenda, the Security Council President approves it, and then the Security Council adopts the agenda. Setting the agenda is a procedural matter where the veto does not apply. Informally, however, the Security Council is unlikely to add an item to the agenda unless the permanent members champion it. The Council usually agrees the agenda in informal meetings without a vote; however, rare situations that evade consensus shed light the process of determining a new agenda item. For example, in 2014, China and Russia opposed adding the situation in North Korea to the agenda. States that championed this built on momentum generated by the extraordinary findings of the Commission of Inquiry on Human Rights in the Democratic People's Republic of Korea (Farrall et al. 2020). The issue was forced to a vote under the leadership of Australia with the support of a diverse group of Council members, and passed despite negative votes from China and Russia (United Nations 2014).

Similarly, in 2006 there was opposition from China on including Myanmar on the Council's agenda. The US led the advocacy towards adding Myanmar to the agenda (Security Council Report 2005). Momentum was built towards this by a series of early briefings under 'other matters' before a contested vote where China, Congo, Qatar, and Russia voted against, and Tanzania abstained, which left a sufficient majority of ten affirmative votes (United Nations 2006).[1] These two examples are rare instances where Security Council members did not reach consensus on new agenda items during informal negotiations, however, they offer insight into how the Council added these agenda items, despite resistance. The

---

[1] A third example is the process by which the Security Council added the situation in Zimbabwe to the agenda in 2005 with the minimum support of nine affirmative votes.

examples highlight the importance of champions inside the Security Council to advocate for new agenda items, and the importance of building momentum via briefings and reports, both of which occurred prior to the Security Council adding Darfur to its agenda.

The Security Council uses the term 'agenda' to mean two different things, which have different political implications (Sievers and Daws 2014: 215). There is the agenda of any given meeting, and there is a running list of previous agenda items called the 'Summary Statement of matters of which the Security Council is seized' or 'Summary Statement' for short, which members also consider to be the 'agenda' of the Security Council. When the Council adds an item to the provisional agenda, and it is adopted as a formal agenda item for a specific meeting, it automatically becomes part of the Summary Statement. Items are only removed from the Summary Statement if there has been no formal meeting under that item for three years, but even then it can be retained at the request of any member of the United Nations (Sievers and Daws 2014: 229–230). Binder and Golub describe this as a process whereby agenda items can become 'sticky' and not easily removed once they are included on the agenda (2020: 419).

The two different uses of the term agenda are important because each type of agenda has different implications. An agenda item for a given topic enables the Security Council to discuss the situation and, if there is enough support, apply the range of tools that are at its disposal, so it is understandable that there may be some hesitation in adding new situations. The Summary Statement has symbolic politics associated with it, whereby the creation of a new agenda item legitimizes the situation as a matter appropriate to the consideration of the Security Council. States recognize the symbolic power that comes from having their favoured issue nominally under consideration, even when the Council is inactive on it (Hurd 2002). For example, maintaining its favoured matter on the Summary Statement has enabled Pakistan to refer to the India–Pakistan question as 'the oldest issue on the UN agenda', and to use this as a legitimation strategy to argue that the matter deserves additional attention (cited in Hurd 2007: 115–116). Once an item is added to the Summary Statement, there has historically been considerable inertia on removing items.[2] In the case of Darfur, media advocacy and domestic salience created the incentive for Western states to champion Darfur in the Security Council.

---

[2] Over the past two decades, there have been efforts to reform this inertia. An item is now automatically scheduled for deletion if it has not been discussed in the past three years unless a United Nations member requests that it remain (which they must do annually to prevent the deletion of an inactive item). For example, this change in procedure has decreased the items on the Summary Statement from 207 in 1993 to sixty-nine items in 2019.

## Availability of Information on Darfur for Decision Makers

Lack of information on the scale and nature of the violence outside Sudan can plausibly explain the lack of Security Council engagement during the first six months of the conflict. It was not until late 2003 that the UN Secretariat and UN member states began to realize the severity of the conflict. This timing is significant because the height of both violence and displacement occurred through 2003 and into 2004, and for the first six months of this violence, there was little outside information or awareness. This lack of information was partly due to the size and remoteness of Darfur, with little presence of outside actors, and partly due to an active campaign by the Sudanese government to restrict movement and access for outsiders.

During the escalation and early stages of the conflict there was almost no UN presence in the region (Traub 2006: 214). The absence of UN personnel on the ground is significant, as the UN privileges UN-derived data as an information source (Jacobsen and Engell 2018: 372–373; Mac Ginty 2017). The United Nations Children's Fund (UNICEF) had a limited presence in Darfur from March 2003, but its movement and activities were severely restricted by the Sudanese Government (UNICEF-DFID 2005). Even with some presence on the ground, UNICEF headquarters in New York reported that it did not become aware of the scale of the Darfur crisis until late 2003 (UNICEF-DFID 2005). Likewise, the UN's Resident and Humanitarian Coordinator for Sudan, Mukesh Kapila, was based in Khartoum from early 2003 but had no staff on the ground in Darfur when the conflict escalated. By mid-2003 he had heard reports of mass violence against civilians, but he reported that at the time 'ours remained a half-formed, patchy understanding' (Kapila and Lewis 2013: 107). Despite being based in Sudan, Kapila himself was unable to visit Darfur until September 2003, which was when the nature and scale of the violence become clearer to him (Cockett 2010: 169–170). With little UN presence in Darfur during the first six months of the crisis, there was very limited information available to the UN Secretariat.

In fact, the first information on the scale and nature of violence in Darfur came from outside the United Nations from NGOs and from a handful of media outlets. Amnesty reported escalating violence in Darfur in February, June, and July of 2003 (Amnesty International 2003a, 2003b, 2003c). In these reports, Amnesty expressed concern over escalating human rights abuses and called for a commission of inquiry to be established. By July 2003, Amnesty described allegations of the Sudanese government bombing civilian areas, and reported that the government had said that it would repress the rebel uprising using military force (Amnesty International 2003a). These attempts to raise the alarm on Darfur went unheeded.

A detailed meta-study of ten key media outlets from different geographical regions and languages found that only four of the ten had issued their first articles on Darfur by mid-2003 (Mody 2010: 341). The Sudanese government actively denied the media access to Darfur. For example, Al Jazeera was one of the first outlets to cover the crisis in Darfur in April 2003, and by December 2003 the government had banned them from Sudan on the grounds of disseminating false-hoods (Mody 2010: 299). Images used in articles on Darfur tended to be taken in refugee camps, where journalists had more access, and tended to present images of women and children in passive situations (Campbell 2007). Across regions and languages, however, news articles and NGO reports were mostly isolated and low in number throughout 2003, showing that the issues in Darfur had not yet gained traction or salience, even when information reached audiences out-side of Sudan. The isolated and limited nature of information—coordinated by Khartoum—meant that there was little capacity to raise the alarm either inside or outside the United Nations during the first six months of extreme violence.

## Opposition from Security Council Members on Engaging with Darfur

The cause of inaction changed six months into the conflict. Through the last few months of 2003, states and departments of the UN Secretariat began to rec-ognize the severity of the conflict and members of the UN Secretariat tried to advocate Security Council engagement. By this stage, however, the UK and the US actively resisted action on Darfur because the north–south negotiations were promising. As Mukesh Kapila explained, in late 2003 he visited various world cap-itals to raise funds for humanitarian aid for Darfur and to raise the alarm. In London and Washington DC, he discovered that the governments had far more intelligence on the escalating crisis than the UN did, and he was shown satellite images of destroyed villages (Kapila and Lewis 2013: 108 and 118). The UK and US were also clear that the Sudanese government were involved in the attacks, given the use of aircraft, but the priority was the Naivasha peace talks between the north and the south of Sudan, and Kapila was repeatedly told 'not to rock the boat' by speaking about Darfur (Kapila and Lewis 2013: 108–119). The barrier was no longer lack of awareness or information on the part of states, but active unwillingness to consider Darfur as an item of the Security Council's agenda.

Médecins Sans Frontières (MSF) concur with this assessment of the situation by late 2003.

> Was the magnitude of the crisis unknown by key players? No. There had been several assessment missions to Darfur between September and December 2003 (UN, USAID, French Ambassador, NGOs), information was available from Dar-furian refugees in Chad. This knowledge did not translate into timely diplomatic activity. The main reason is that the international community was deliberately ignoring the Darfur crisis in order to preserve the north–south peace process.
>
> (Weissman 2007)

Towards the end of 2003—six months into the conflict—there was knowledge of the crisis in Darfur in key capitals. There was growing understanding of the scope and nature of the violence and of Khartoum's involvement. Political considerations now hampered Security Council action, rather than lack of awareness. This was the first real failure to engage with the crisis in Darfur. We cannot know what might have happened differently with a strong diplomatic response at this early stage of the conflict, but it may have been able to change Khartoum's calculations, as some have argued (Cockett 2010: 199; Traub 2010). Competing priorities in Sudan meant that the international community implicitly gave Khartoum a free reign to orchestrate violence in Darfur.

The UN Secretariat has a key role in providing information and advice to the Security Council due to its legitimated authority, as discussed in Chapter 1. Divisions between Secretariat departments diluted this message, however, and gave states a range of arguments to choose from, where they could choose the argument that most closely aligned with their interests. This division was between the Office for the Coordination of Humanitarian Affairs (OCHA), which coordinates humanitarian actors and has an advocacy role, and the UN's Department of Political Affairs (UN DPA)[3] which provides internal political guidance for the UN Secretary-General and other UN departments and agencies.[4] These two departments disagreed on how to interpret the situation on the ground, what actions the UN should take, and whether there was any need for advocacy on this issue. OCHA staff, with a humanitarian remit, interpreted the situation in Darfur as a political crisis requiring an urgent political response from the UN Security Council, while staff from the UN DPA, with a political remit, viewed it as tribal violence rather than as exceptional government-sponsored violence.

The UN DPA was established as the 'primary conduit to the UN Secretary-General in relation to countries at risk of conflict' (Bellamy 2011: 133; see also McLoughlin et al. 2023). However, in this case, the UN DPA advised the UN Secretary-General not to engage with Darfur and argued that they should not represent the situation in political terms. OCHA's Resident and Humanitarian Coordinator in Sudan, Mukesh Kapila, and his boss Jan Egeland recognized the exceptional nature of the violence and its ethnic dimension. They met with representatives from the UK, the US, and Norway, who made up the troika working on the north–south peace negotiations. The message they received was consistent; states were aware of the situation in Darfur from their own intelligence sources

---

[3] I use the abbreviation 'UN DPA' to refer to the UN's former Department of Political Affairs over the standard abbreviation 'DPA' for clarity, as I later discuss the Darfur Peace Agreement, which also uses the abbreviation 'DPA'.

[4] For more background see, respectively http://www.unocha.org/about-us/who-we-are and http://www.un.org/undpa/overview. Concordant with their actions, advocacy was formally part of OCHA's role during the 2003–2004 period, see https://docs.unocha.org/sites/dms/Documents/OCHA2004.pdf

but wanted a breakthrough in the north–south negotiations before they broached Darfur (Egeland 2008; Kapila 2006). Troika members thought that criticizing government actions in Darfur would jeopardize the peace talks.

In contrast, the UN DPA interpreted the situation in Darfur as a humanitarian crisis that required humanitarian aid. Kieran Prendergast, who was then Under-Secretary-General for Political Affairs, says that the first time he became aware of the conflict in Darfur was in late 2003, but he said that deaths at this time in Darfur were from disease and lack of water, rather than from attacks (Frontline 2007a).[5] Even when he became aware of attacks, Prendergast interpreted the situation as unexceptional violence rather than as coordinated and systematic violence by the government (Frontline 2007b; Kapila and Lewis 2013: 114).[6] For Prendergast, the ongoing north–south negotiations were thought to be a precursor to broader stability in Sudan (Frontline 2007a). While arguing for quiet diplomacy, Prendergast also criticized OCHA's advocacy on Darfur as a political problem on the grounds that it was outside of their 'humanitarian' mandate (Author Interview 2012; Kapila and Lewis 2013: 187). This meant that two different arguments were emerging from the Secretariat—one that interpreted the situation as an urgent political crisis and one that interpreted it as an unremarkable humanitarian crisis. This gave states licence to choose which argument to align with, as both carried the 'legitimacy' of the Secretariat's authority.

Divided perspectives within the UN Secretariat also limited the capacity of the Secretariat to lobby Security Council members. While OCHA lobbied members of the UN Security Council to include Darfur on the Council's agenda, the UN DPA did not think this was necessary. Indeed, Egeland and Kapila both describe how their reports and requests for political support from the UN DPA went unheeded (Kapila 2006; Kapila and Lewis 2013; Traub 2010).[7] There are also potentially more pernicious effects of this disputed framing. The depoliticized framing used by the UN DPA aligned with the framing initially used by the UK and the US to justify their exclusive focus on the conflict in southern

---

[5] Current knowledge about mortality patterns in Darfur contradicts this understanding (Degomme, Guha-Sapir 2010). There was also information available to Prendergast in 2003 that contradicted his interpretation. Kapila expressed serious concern to Prendergast in person about the escalating situation in Darfur around August 2003, including the use of aircraft, which suggested Khartoum's involvement (Kapila and Lewis 2013). Likewise, a memo from Kapila to Prendergast on 18 December 2003 said that the conflict had affected one million people. It also said that 70,000 had fled into Chad and that OCHA's offices were receiving daily reports of human rights violations (Frontline 2007c). In this memo, Kapila described the conflict in Darfur as 'war', with people fleeing regular human rights abuses, which contrasts with Prendergast's framing of the situation as a humanitarian crisis which simply required humanitarian aid.

[6] There are parallels here with the frame of 'civil war' used by the UN Secretariat during the Rwandan genocide: a misrepresentation that led to a reluctance to act (Barnett 2002; Barnett and Finnemore 2004; Melvern 2001).

[7] Some authors suggest that UN staff may not have taken Kapila particularly seriously when he tried to raise attention on Darfur, as he was considered by some to be an alarmist (Cockett 2010; Traub 2006).

Sudan (Srinivasan 2013). The UN DPA's representation of Darfur as a humanitarian crisis helped to legitimize a depoliticized approach. For example, similar to Prendergast, Charles Snyder from the US State Department interpreted the situation as 'a standard African civil war'; normalized violence rather than exceptional mass atrocity crimes (Cockett 2010: 196). Likewise, statements given by the British Foreign Office a month apart were almost identical in wording, except that the first statement referred to 'armed groups' and 'armed fighters', while in the later statement this had been replaced with 'tribal groups', downgrading the political nature of the violence (Srinivasan 2013: 31). The US and UK took steps to depoliticize their representation of the Darfur conflict in the early stages, thus justifying their focus on the north–south peace talks. The position taken by the UN DPA—and divisions within the Secretariat—bolstered the approach taken by Western states to focus on the north–south conflict to the exclusion of Darfur at this time.

## Incremental Steps towards Darfur as an Agenda Item

### Media and the Tenth Anniversary of the Rwandan Genocide

Mukesh Kapila linked the situation in Darfur via analogy to the Rwandan genocide, after reaching a point of frustration with member state inaction, which has aptly been described as a 'bridge-building metaphor' by Lanz (2020: 47). Statements made by Kapila and others to publicly link these cases shaped the arguments that Western states could make which would be regarded as legitimate by domestic audiences. Kapila made a statement to the media on 19 March 2004 where he described the situation in clear political terms as 'ethnic cleansing' and said that 'the only difference between Rwanda and Darfur is the numbers involved of dead, tortured and raped' (AFP 2004). Kapila's statement was the first public representation of the crisis as political and as having similarities with atrocities on the scale of the Rwandan genocide, where 800,000 people were killed in a hundred days. The statement triggered a dramatic increase in the international media's interest in Darfur (Grzyb 2010; Kapila and Lewis 2013). This domestic and media interest made it difficult for Western states to continue to sideline Darfur in favour of the north–south negotiations. To be clear, by late 2003 Western states had extensive intelligence demonstrating the nature of violence in Darfur. Western governments knew that civilians were being targeted with violence and that the perpetrators had aviation capacity, which logically restricted this to the government, or to government-supported actors. It was not the nature of the violence that changed which arguments they could legitimately make, but the public nature of the violence.

Kapila's statement gained traction over earlier statements from staff of the UN Secretariat. While Kapila has been described as the 'whistle-blower' on Darfur,

Egeland and Annan had previously spoken to the media about Darfur in December 2003. Egeland said that 'the humanitarian situation in Darfur has quickly become one of the worst in the world' (UN News Centre 2003) and Annan expressed alarm 'at the rapidly deteriorating humanitarian situation in the Darfur region' (United Nations 2003). There were two reasons, however, why Kapila's statement gained traction while the earlier statements did not. First, while Egeland and Annan focused on the humanitarian aspects of the crisis, Kapila's statement stressed the political nature of the violence. This framing gave greater salience as an exceptional situation that required urgent attention. Simply put, Kapila's statement was 'blunter' and less diplomatic than the other Secretariat members (Flint and de Waal 2008: 179). The second reason was timing: Kapila made his statement only weeks before the tenth anniversary of the Rwandan genocide. Media outlets were already preparing to cover the anniversary, and the link to Darfur gave them a new angle. The statement from Kapila has therefore been accurately described as the 'spark' which triggered a rapid increase in media on Darfur in the United States (Grzyb 2010: 81; Stedjan and Thomas-Jensen 2010: 159).

The uptick in media interest did not reflect increased violence in Darfur; in fact, it was the inverse. A search of the online news archive Factiva showed that there were less than 500 headlines globally featuring the word 'Darfur' in the twelve months leading up to Kapila's statement—the period which corresponded with the highest levels of violence against civilians—and almost 10,000 headlines in the twelve months following Kapila's statement. As Figure 6 shows, there was a rapid rise in media interest in Darfur in the early months of 2004, with a peak in media interest in mid-2004.[8] The level of media interest was not proportionate with the scale of the conflict in Darfur. A comparison between Figure 4 and Figure 6 shows that while the height of violence occurred in late 2003 and early 2004, media interest grew from March 2004. This pattern was common across media outlets, regions, and languages, with isolated or infrequent articles in 2003 and a rapid growth in media interest in early 2004 (Mody 2010). With Kapila's statement, and links to the Rwandan genocide, Darfur became a major news story, despite decreased violence.

There was nothing inevitable about linking the situation in Darfur to the Rwandan genocide, which could have been interpreted via analogy with another case (Hamilton 2011: 32). Indeed, for some, the relevant frame for the Darfur crisis was not Rwanda, but the 2003 Iraq War, connected to concerns that the calls for intervention on humanitarian grounds represented a cover for Western interests (for discussion see Verhoeven, de Oliveira and Jaganathan 2015). However, it was the analogy to the Rwandan genocide that the media framed as legitimate, and this connection had a significant effect on Darfur advocacy, not only in 2004 but also

---

[8] The subsequent spikes in 2006 and 2007 map onto the international debates around UN peacekeeping. See Chapter 7 for more on this.

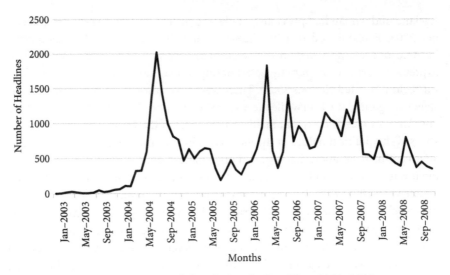

**Figure 6** Newspaper Articles with 'Darfur' in the Headline, 2003–2008
*Source:* Compiled from the Factiva database

for many years afterwards, reflected in ongoing debates over whether the situation in Darfur constituted genocide. In relation to agenda-setting, framing Darfur as analogous to the Rwandan genocide, with the anniversary approaching, significantly elevated Darfur within the media and tied these two cases together in public perception (Brunk 2008; Gifkins 2016; Mamdani 2009). Media outlets picked up on the analogy between the Rwandan genocide and Darfur, including Samantha Power, author of 'A Problem from Hell: America and the Age of Genocide', who published a *New York Times* op-ed titled 'Remember Rwanda, but Take Action in Sudan' (2004b). African leaders also drew on the parallels between Rwanda and Darfur to make the case for action in Darfur (Verhoeven, de Oliveira and Jaganathan 2015). Even though the analogy between Rwanda and Darfur was far from inevitable, it generated salience and became the legitimated frame for the situation in Darfur.

## Movement within the UN Security Council

By April 2004, over a year after the conflict in Darfur escalated, OCHA was still advocating for access to the Security Council against opposition from the UN DPA. Egeland tried to enlist Prendergast's help in gaining access to the Council, but reflecting on this period he recalled 'It was clear that the Department of Political Affairs would not help me with this' (Traub 2006: 218). In April, however, Germany held the presidency of the Council and worked with the British and

French to invite Egeland to brief the Council (Traub 2006: 220). There was still too much resistance at this time to create an agenda item for Darfur, so this briefing was not a formal meeting of the Council, but was an informal 'consultation of the whole' (United Nations 2004a). 'Consultations of the whole' are not formal Security Council meetings, but 'informal gatherings of all fifteen members' which do not transfer onto the Summary Statement (Sievers and Daws 2014: 65).[9] As such, it was a slightly less contentious meeting format.

Given that informal consultations are not public, Egeland took the opportunity to hold a UN press conference to share his report publicly. Egeland spoke for half an hour, described the situation as 'ethnic cleansing', and said that the Janjaweed militia were specifically targeting Black African communities in Darfur and that the government was not trying to stop the attacks (United Nations 2004c). Egeland was explicit that the violence was ethnically motivated and suggested that it was at least condoned by the Sudanese government, aligning with Kapila's earlier framing of Darfur as a political crisis. Again showing the divisions between OCHA and the UN DPA, Prendergast recalled advising Egeland not to describe the situation as 'ethnic cleansing' at this briefing (Frontline 2007a). After many months of lobbying, OCHA managed to brief the Security Council on the political dimensions of the conflict in Darfur, despite resistance from the UN DPA.

In response to Egeland's briefing, the UN Security Council issued a statement to the press on Darfur, which is the least formal output from the Security Council. The statement expressed 'deep concern about the massive humanitarian crisis' and welcomed the peace talks taking place in N'Djamena on Darfur (United Nations 2004d). Press statements require the Council to speak with one voice, and the positions taken by Chinese, Algerian, and Pakistani diplomats meant that the language was weak (Egeland 2008: 90). Opposition to describing the conflict in political terms or criticizing the Sudanese government would later become a regular division within Council negotiations on Darfur, which this early example foreshadowed. This lowest-common-denominator statement framed Darfur as a purely humanitarian crisis, even though Egeland's briefing clearly argued that the crisis was political in origin.

The peace talks in N'Djamena referred to in the Council's statement were concluded the following week. There was intense pressure on the negotiations from the US for a 'quick fix', but neither party to the conflict respected the ceasefire agreement (Flint and de Waal 2008: 174–175). International attention remained on the north–south peace negotiations and it was thought that a quick peace negotiation might resolve the situation in Darfur so that the focus could remain on Sudan's other conflict. The rushed mediation effort, led by the African Union and Chad, resulted in the N'Djamena Humanitarian Ceasefire Agreement. This

---

[9] For a detailed account of all types of Security Council meetings and consultations see Sievers and Daws (2014: 19–97).

agreement had two major flaws, even though it was signed by all major parties to the conflict, the government of Sudan and the JEM and SLM/A. First, an additional sentence was added after all parties had signed, stating that rebel groups would be assembled in designated areas, which Khartoum argues was the official version. The rebels completely disagreed with this (Toga 2007: 217). Second, the agreement did not contain any maps to show the location of armed forces so infringements of the ceasefire could be monitored (de Waal 2007: 377). Almost immediately after the ceasefire was signed, there were reports that it was being breached by Janjaweed members attacking civilians (Human Rights Watch 2004b: 52). Symptomatic of the international focus on the north–south negotiations, the rushed ceasefire agreement foreshadowed the later negotiations towards the Darfur Peace Agreement, which similarly prioritized achieving 'a peace deal' over building momentum for a sustainable peace.

## Shifting Priorities within UN Security Council Member States on Darfur: The US and the UK

After months of actively resisting Darfur as an agenda item, priorities changed for the US government. As late as February 2004, the US refused to allow Egeland to brief the Council on Darfur (Egeland 2008: 90). Yet by May 2004, Human Rights Watch (HRW) describe the US as the government which had taken the strongest public stance on Darfur (Human Rights Watch 2004b: 58). This shift requires explanation. In early 2004, the US government's top two priorities in Sudan were the ongoing north–south peace talks and counterterrorism cooperation (Stedjan and Thomas-Jensen 2010). The conflict in southern Sudan had strong resonance in the US for two key reasons (Cockett 2010: 152–160). First, Christians were predominately targeted in the violence and Christianity is deeply embedded in US domestic politics. Second, slavery was ongoing in Sudan through the 1980s and 1990s, which had salience in US domestic politics, and US civil society groups opposed to slavery in Sudan had developed. These two elements meant that Sudan—and in particular the conflict in southern Sudan—resonated within lobby groups in the US more than it did elsewhere (Cockett 2010: 155).

The resonance of the north–south conflict, combined with Sudan's counterterrorism relationship with the US, meant that Sudan was a key foreign policy issue within the US. Indeed, when former US envoy to Sudan John Danforth was asked about the relative importance of Sudan within US foreign policy, he replied, 'it was a very, very high priority, it was a remarkably high priority' (Keane 2005). A key reason for this was that US President Bush was elected with strong support from evangelical Christians, and one of his election pledges was to help Sudanese Christians (Stedjan and Thomas-Jensen 2010: 160–161). By December 2003 Bush was so confident that a north–south peace deal was imminent that two seats were reserved at the US State of the Union address, for Sudanese President Omar al-Bashir and South Sudanese leader John Garang, to celebrate President Bush's imminent success in negotiating a peace agreement (Power 2004a). With the end to one of Africa's longest civil wars in sight, Washington was highly focused on achieving that goal and consequently sidelined Darfur in late 2003 and early 2004. Hamilton describes 'cognitive dissonance' at work in the US government, where a tension between brokering peace in the south of Sudan or taking a political approach to Darfur led officials to discount one set of information, namely the political nature of the Darfur crisis (2011: 29). As such, in early 2004, Darfur could, at most, be considered a third priority within US foreign policy on Sudan, which was not high enough to compete with the plethora of other foreign policy priorities in different countries and regions, let alone with other priorities in Sudan.

However, the connection between the US and southern Sudan meant that there were ready-made lobby groups which were interested in Sudan and became interested in Darfur. Evangelical Christians and the Congressional Black Caucus quickly became engaged with advocacy on Darfur, drawing on prior connections with southern Sudan (Heinze 2007). Shifts began to appear in statements made by President Bush in April 2004. Up to this point, the US had told Khartoum that it would normalize relations once the north–south peace agreement was finalized. Normalizing relations was a goal for both states. However, on the anniversary of the Rwandan genocide, President Bush issued two statements: one which marked the anniversary of the Rwandan genocide and one which condemned atrocities in Darfur—during which it was suggested that normalizing relations with Khartoum might be contingent on it at least allowing humanitarian access in Darfur (Bush 2004a, 2004b). Media interest and the anniversary of the Rwandan genocide also triggered the beginning of a civil society movement on Darfur specifically.

The US already had a strong civil society movement on southern Sudan, and through the first half of 2004 an additional civil society movement on Darfur was growing (Hamilton 2011). It became untenable for the US government to continue to ignore Darfur in favour of the north–south negotiations because Darfur had a growing media profile and pre-existing caucus and civil society groups focused on Sudan. To seek legitimacy for the arguments it was making, and by extension for itself as an actor, the US government chose to actively engage with the crisis in Darfur. US intelligence on the escalating crisis did not drive the shift towards US advocacy; instead, it was driven by OCHA's media advocacy, the anniversary of the Rwandan genocide, and domestic civil society interest in Sudan. Media and civil society interest in Darfur increased the legitimacy costs for the US to normalize relations with Sudan while sidelining Darfur.

Like the US, the UK's shift in stance on Darfur in the Security Council followed the same trajectory in early 2004. When Kapila travelled to London around August 2003 and January 2004 to raise awareness on Darfur, he found that government officials were already aware of the situation on the ground in Darfur (Kapila and Lewis 2013: 108–110 and 182–185). The UK told Kapila that the north–south peace process was too important to jeopardize with political action on Darfur, and the UK responded with funding for humanitarian aid instead. Indeed, the British government presented a depoliticized account of the Darfur conflict in the early stages—which did not match with their internal understanding—to legitimize their continued focus on the north–south peace talks over Darfur (Srinivasan 2013). As it had in the US, Kapila's statement linking Darfur and Rwanda triggered a rapid increase in British news stories on Darfur (Campbell 2007). British foreign policy, under Prime Minister Tony Blair, had a strong focus on Africa, and civil society pressured the UK government to take a stronger stand on Darfur (Williams 2010: 201–203). Darfur did not become as high a foreign policy priority for the UK as it did for the US, however, as the prior connections to slavery and persecution of Christians had a particular salience in US politics, which created stronger lobby groups on Sudan (Cockett 2010). Domestic interest in Darfur shifted the UK's position on engaging with this conflict.

The legitimacy costs for both the UK and the US changed in the early months of 2004, when they shifted from actively resisting Council engagement with Darfur to becoming the Council's most vocal advocates on Darfur. This was a function of media engagement and domestic salience, rather than simply a function of the government's intelligence on state-sponsored violence in Darfur. Concordant with Boulden's argument, domestic salience and civil society activism incentivized P3 interest in championing Darfur in the Security Council (2006). Without this critical ingredient of publicity and domestic salience, it is likely that Security Council members would have taken longer to engage with Darfur. As such, an external legitimation practice—legitimation of the actor—led to the shift in position for the US and the UK. To present a consistent argument to domestic audiences on the protection of civilians in Sudan, both states had to shift their position and become advocates for Darfur.

## Darfur as an Agenda Item in the Security Council

Now that there were permanent members championing Darfur in the Security Council, it was possible to build momentum towards an agenda item. Egeland's briefing to the Council on Darfur established a precedent and the Security Council began to receive regular briefings; however, in these early stages the Council only discussed Darfur at informal 'consultations of the whole', rather than as an item with its own listing on the agenda. For example, on 7 May 2004, James Morris, the Executive Director of the UN's World Food Programme, and Bertrand Ramcharan, the Acting UN High Commissioner for Human Rights, briefed the Council. They told the Security Council that Khartoum was sponsoring the Janjaweed militia in a 'reign of terror' in Darfur, stressing the political dimensions of the crisis (ICG 2004: 11).

Moving on from 'UN only' briefings, on 24 May 2004, the UN Security Council held its first 'Arria formula' meeting on Darfur, initiated by Germany, which enabled NGOs to brief the Council rather than only UN entities. This Arria formula meeting included presentations from MSF, International Crisis Group (ICG), HRW, and the Deputy Permanent Representative of Sudan (Sievers and Daws 2014: 84). Ton Koene from MSF briefed the Security Council and stated that the government of Sudan and Janjaweed militia were perpetrating widespread violence against the civilian population and that the government of Sudan was deliberately restricting the access of humanitarian organizations (Koene 2004). Similarly, HRW particularly highlighted the close relationship between Khartoum and the Janjaweed militia (Human Rights Watch 2004a). Regular UN and NGO briefings to the Council became a key tool to build momentum for an agenda item on Darfur.

By May 2004, Western states had begun to champion Darfur, and a new dynamic emerged within the Security Council. While Western states had earlier opposed any engagement with Darfur, they were now critical of Khartoum's human rights violations in Darfur. This in turn created tensions with states in the Council that were allies of Sudan. Throughout May, Khartoum lobbied members of the Security Council to resist the inclusion of Darfur on the agenda, with recognition of the significance that an agenda item holds both politically and symbolically. At this time four Council members—Pakistan, Russia, China, and Algeria—tried to prevent the new agenda item (Arieff 2004). However, Darfur now had the critical ingredients for the creation of an agenda item: champions within the Council and momentum via regular briefings. Even though two permanent members opposed the creation of an agenda item, there is no veto on procedural items, and the P3 leverage their institutional power to dominate the agenda-setting process. The states advocating the inclusion of Darfur on the agenda were the US, the UK, Germany, and France (Author Interview 2011). Unlike the examples highlighted earlier, the Security Council agreed to add Darfur to the agenda during informal consultations, so it did not hold a vote. On 25 May, the day after the Arria formula meeting, the Council adopted a new agenda item on Darfur, with no objections (United Nations 2004b).[10] By this time, sufficient momentum had developed through regular briefings to the Council from UN staff and NGOs.

[10] The Security Council initially added Darfur as an agenda item titled 'Letter dated 25 May 2004 from the Permanent Representative of the Sudan to the United Nations addressed to the President of the Security Council'. The Council held only one formal meeting under this agenda item, and an additional twenty 'consultations of the whole'. At this stage, there were two separate agenda items for Sudan: the above item for Darfur and one to discuss the north–south conflict called 'Letter dated 2 October 2003 from the Permanent Representative of the Sudan to the United Nations addressed to the President of the Security Council'. This division reflects the fragmented approach taken by the Council, leading it to respond to Sudan's two conflicts in a disconnected way. The Council later joined the two

Media interest changed the arguments that Western permanent members could make that would gain legitimacy with domestic audiences, and this facilitated the addition of Darfur to the Security Council's agenda. The UK and the US initially chose to sideline Darfur in favour of the north–south negotiations, even when they had intelligence on the scale and nature of the violence, until this became untenable with increased media and domestic salience. The divergence in perspectives between OCHA and the UN DPA initially allowed Western states to choose which interpretation of the situation best suited their political priorities, until OCHA's position became the dominant frame for Darfur. Analogy to the horrors of the Rwandan genocide enabled the salience and rapid escalation of public interest in Darfur. Unlike subsequent chapters in this book, the agenda-setting process happened largely outside the Security Council, primarily through lobbying from the UN Secretariat, and when that failed, through media attention to shift the calculations on legitimate arguments from Western states. The US became a vocal advocate on Darfur; however, Darfur remained at most the third policy priority for the US within Sudan, behind counterterrorism and the north–south peace talks. Strong rhetoric on Darfur then became a substitute for action (Heinze 2007; Vik 2015; Williams 2010). Advocacy by OCHA eventually created the conditions for Security Council engagement, however, this was more than a year after the conflict escalated and after the highest rates of violence had passed.

## Conclusion

Darfur became part of the Security Council's agenda after the US and the UK shifted positions in late 2003 and early 2004. This change came about via a circuitous route whereby OCHA staff tried to lobby Western states directly, but when this failed, Kapila approached the media, which in turn generated domestic salience and changed the arguments that the US and the UK could legitimately make in public. These states were aware of the scale and nature of the violence against civilians by late 2003, as Kapila found when he visited capitals and senior officials showed him extensive surveillance of destroyed villages. As such, it was not simply knowledge of the violence that changed the calculation of these states, but public awareness of the violence. While Western states went on to become prominent and vocal advocates on Darfur, the initial response was to sideline Darfur in favour of the north–south peace process, which complicates the narrative of Western states as leading advocates on Darfur. As is evident in Boulden's framing on which situations become part of the Security Council's agenda, public awareness and advocacy can be sufficient to cause a shift from lack of engagement

agenda items into a single agenda item, listed as 'Report of the Secretary-General on the Sudan'. For full details, see United Nations (2004e).

by permanent members to engagement (2006). A key factor to enable creation of a new agenda item is P3 members championing the issue and building momentum via reports and briefings, as occurred in this case. It was external legitimation practices—whereby states seek legitimacy for their arguments and by extension themselves as actors—which facilitated the shift of the UK and the US towards becoming vocal advocates on Darfur.

This chapter is unlike the other case study chapters in this book in that the subsequent chapters focus on negotiations inside the Security Council, while much of the process of establishing Darfur as an agenda item happened outside the Council with advocacy from the UN Secretariat, NGOs, and the media. Activism from the UN Secretariat was critical in generating domestic salience for the crisis in Darfur; however, divisions between OCHA and the UN DPA meant that there were opposing arguments that had the legitimated authority of the Secretariat. Western states initially chose the argument that best suited their purposes of focusing on the north–south negotiations. However, once Darfur became a significant media story—admittedly after the height of the violence had passed—the US and the UK changed their position to one that domestic audiences would be more likely to regard as legitimate. The shift in positions which enabled Darfur to become an agenda item were brought about by changed perceptions of which arguments could be legitimately made and the external legitimation practices of states seeking legitimacy in the eyes of audiences that were important to them.

## Bibliography

AFP. 2004. 'West Sudan's Darfur Conflict "World's Greatest Humanitarian Crisis"'. *Sudan Tribune*, 19 March.

Allen, Susan Hannah and Amy Yuen. 2022. *Bargaining in the UN Security Council: Setting the Global Agenda*. Oxford: Oxford University Press.

Amnesty International. 2003a. *Sudan: Empty Promises? Human Rights Violations in Government Controlled Areas*. Available at https://www.amnesty.org/en/documents/afr54/036/2003/en/.

Amnesty International. 2003b. *Sudan: Looming Crisis in Darfur*. Available at https://www.amnesty.org/en/documents/afr54/041/2003/en/.

Amnesty International. 2003c. *Sudan: Urgent Call for Commission of Inquiry in Darfur as Situation Deteriorates*. Available at http://www.amnesty.org.uk/press-releases/sudan-urgent-call-commission-inquiry-darfur-situation-deteriorates.

Arieff, Irwin. 2004. 'UN Tiptoes around Khartoum in Coping with Darfur'. *Sudan Tribune*, 28 May.

Author Interview. 2011. *Author Interview with Senior P5 member*. 20 September, Cambridge.

Author Interview. 2012. *Author Interview with UN Source on Darfur*. 11 August, telephone interview.

Barnett, Michael N. 2002. *Eyewitness to a Genocide: The United Nations and Rwanda*. New York: Cornell University Press.

Barnett, Michael N. and Martha Finnemore. 2004. *Rules for the World: International Organizations in Global Politics*. Ithaca and London: Cornell University Press.

Bellamy, Alex J. 2011. *Global Politics and the Responsibility to Protect: From Words to Deeds*. Global Politics and the Responsibility to Protect. New York: Routledge.

Binder, Martin and Jonathan Golub. 2020. 'Civil Conflict and Agenda-Setting Speed in the United Nations Security Council'. *International Studies Quarterly* 64(2): 419–430.

Boulden, Jane. 2006. 'Double Standards, Distance and Disengagement: Collective Legitimization in the Post-Cold War Security Council'. *Security Dialogue* 37(3): 409–423.

Brunk, Darren. 2008. 'Dissecting Darfur: Anatomy of a Genocide Debate'. *International Relations* 22(1): 25–44.

Bush, George W. 2004a. *1994 Rwanda Genocide*. Washington DC, 7 April: US Department of State. Available at http://2001-2009.state.gov/s/wci/us_releases/rm/31350.htm.

Bush, George W. 2004b. *President Condemns Atrocities in Sudan*. Washington DC, 7 April: US Department of State. Available at http://2001-2009.state.gov/p/af/rls/rm/31206.htm.

Campbell, David. 2007. 'Geopolitics and Visuality: Sighting the Darfur Conflict'. *Political Geography* 26: 357–382.

Cockett, Richard. 2010. *Sudan: Darfur and the Failing of an African State*. New Haven and London: Yale University Press.

Degomme, Olivier and Debarati Guha-Sapir. 2010. 'Patterns of Mortality Rates in Darfur Conflict'. *Lancet* 375: 294–300.

de Waal, Alex. 2007. 'Darfur's Elusive Peace'. In *War in Darfur and the Search for Peace*, ed. Alex de Waal. Global Equity Initiative, London: Harvard University and Justice Africa.

Egeland, Jan. 2008. *A Billion Lives: An Eyewitness Report from the Frontlines of Humanity*. New York: Simon & Schuster.

Farrall, Jeremy, Marie-Eve Loiselle, Christopher Michaelsen, Jochen Prantl, and Jeni Whalen. 2020. 'Elected Member Influence in the United Nations Security Council'. *Leiden Journal of International Law* 33(1): 101–115.

Flint, Julie and Alex de Waal. 2008. *Darfur: A New History of a Long War*. African Arguments. London and New York: Zed Books.

Frederking, Brian and Christopher Patane. 2017. 'Legitimacy and the UN Security Council Agenda'. *PS: Political Science and Politics* 50(2): 347–353.

Frontline. 2007. *Mukesh Kapila's Memos to the United Nations*. Available at http://www.pbs.org/wgbh/pages/frontline/darfur/etc/memo.html.

Frontline. 2007a. *Interview Kieran Prendergast*. Available at http://www.pbs.org/wgbh/pages/frontline/darfur/interviews/prendergast.html.

Frontline. 2007b. *Interview James Traub*. Available at http://www.pbs.org/wgbh/pages/frontline/darfur/interviews/traub.html.

Gifkins, Jess. 2016. 'Naming and Framing: Darfur, the Genocide Debate, and the Responsibility to Protect'. In *The United Nations and Genocide*, ed. Deborah Mayerson. Basingstoke: Palgrave Macmillan.

Grzyb, Amanda F. 2010. 'Media Coverage, Activism, and Creating Public Will for Intervention in Rwanda and Darfur'. In *The World and Darfur: International Response to Crimes against Humanity in Western Sudan*, 2nd ed., ed. Amanda F. Grzyb. Montreal and Kingston: McGill-Queen's University Press.

Hamilton, Rebecca. 2011. *Fighting for Darfur: Public Action and the Struggle to Stop Genocide*. New York: Palgrave Macmillan.

Heinze, Eric A. 2007. 'The Rhetoric of Genocide in U.S. Foreign Policy: Rwanda and Darfur Compared'. *Political Science Quarterly* 122(3): 359–383.

Human Rights Watch. 2004a. *Addressing Crimes Against Humanity and Ethnic Cleansing in Darfur, Sudan: A Briefing Paper to the UN Security Council*. New York Human Rights Watch. Available at http://www.hrw.org/legacy/english/docs/2004/05/24/darfur8811_txt.htm.

Human Rights Watch. 2004b. *Darfur Destroyed: Ethnic Cleansing by Government and Militia Forces in Western Sudan*. Available at http://www.hrw.org/sites/default/files/reports/sudan0504full.pdf.

Hurd, Ian. 2002. 'Legitimacy, Power, and the Symbolic Life of the UN Security Council'. *Global Governance* 8(1): 35–51.

Hurd, Ian. 2007. *After Anarchy: Legitimacy and Power in the United Nations Security Council*. Princeton: Princeton University Press.

ICG. 2004. *Sudan: Now or Never in Darfur*. International Crisis Group.

Jacobsen, Katja Lindskov and Troels Gauslå Engell. 2018. 'Conflict Prevention as Pragmatic Response to a Twofold Crisis: Liberal Interventionism and Burundi'. *International Affairs* 94(2): 363–380.

Kapila, Mukesh. 2006. 'Why the International Community Failed Darfur'. In *Darfur: The Responsibility to Protect*, eds. David and Alexander Ramsbotham. Mepham: Institute for Public Policy Research.

Kapila, Mukesh and Damien Lewis. 2013. *Against a Tide of Evil: How One Man Became the Whistleblower to the First Mass Murder of the Twenty-First Century*. Edinburgh and London: Mainstream Publishing.

Keane, Fergal. 2005. *John Danforth Interview Transcript*. BBC. Available at http://news.bbc.co.uk/2/hi/programmes/panorama/4647211.stm.

Koene, Ton. 2004. *The Humanitarian Situation in Darfur, Sudan*. Médecins Sans Frontieres. Available at http://www.doctorswithoutborders.org/news-stories/speechopen-letter/humanitarian-situation-darfur-sudan-0.

Lanz, David. 2020. *The Responsibility to Protect in Darfur: From Forgotten Conflict to Global Cause and Back*. Global Politics and the Responsibility to Protect. Abingdon and New York: Routledge.

Mac Ginty, Roger. 2017. 'Peacekeeping and Data'. *International Peacekeeping* 24(5): 695–705.

Mamdani, Mahmood. 2009. *Saviors and Survivors: Darfur, Politics and the War on Terror*. New York: Pantheon Books.

McLoughlin, Stephen, Jess Gifkins and Alex J. Bellamy. 2023. 'The Evolution of Mass Atrocity Early Warning in the UN Secretariat: Fit for Purpose?'. *International Peacekeeping* OnlineFirst.

Melvern, Linda. 2001. 'The Security Council: Behind the Scenes'. *International Affairs* 77(1): 101–111.

Mody, Bella. 2010. *The Geopolitics of Representation in Foreign News: Explaining Darfur*. Plymouth: Lexington Books.

Power, Samantha. 2004a. *Dying in Darfur: Can the Ethnic Cleansing in Sudan be Stopped? The New Yorker*, 22 August. Available at http://www.newyorker.com/archive/2004/08/30/040830fa_fact1.

Power, Samantha. 2004b. 'Remember Rwanda, but Take Action in Sudan'. *The New York Times*, 6 April.

Security Council Report. 2005. *Update Report: Myanmar.* Available at http://www.securitycouncilreport.org/atf/cf/%7B65BFCF9B-6D27-4E9C-8CD3-CF6E4FF96FF9%7D/Update%20Report%2015%20December%202005_Myanmar.pdf.

Sievers, Loraine and Sam Daws. 2014. *The Procedure of the UN Security Council,* 4th ed. Oxford: Oxford University Press.

Srinivasan, Sharath. 2013. 'Negotiating Violence: Sudan's Peacemakers and the War in Darfur'. *African Affairs* 113(450): 24–44.

Stedjan, Scott and Colin Thomas-Jensen. 2010. 'The United States'. In *The International Politics of Mass Atrocities: The Case of Darfur,* eds. Paul D. Williams and David R. Black. London and New York: Routledge.

Toga, Dawit. 2007. 'The African Union Mediation and the Abuja Peace Talks'. In *War in Darfur and the Search for Peace,* ed. Alex de Waal. Global Equity Initiative, London: Harvard University and Justice Africa.

Traub, James. 2006. *The Best Intentions: Kofi Annan and the UN in the Era of American World Power.* New York: Farrar, Straus and Giroux.

Traub, James. 2010. *Unwilling and Unable: The Failed Response to the Atrocities in Darfur.* New York. Available at http://www.responsibilitytoprotect.org/GCR2P_UnwillingandUnableTheFailedResponsetotheAtrocitiesinDarfur.pdf.

UN News Centre. 2003. *Humanitarian and Security Situations in Western Sudan Reach New Lows, UN Agency Says.* Available at http://www.un.org/apps/news/storyAr.asp?NewsID=9094&Cr=sudan&Cr1=.

UNICEF-DFID. 2005. *Global: Joint UNICEF-DFID Evaluation of UNICEF Preparedness and Early Response to the Darfur Emergency.* Available at http://www.unicef.org/evaldatabase/index_31130.html.

United Nations. 2003. *Secretary-General Alarmed by Deteriorating Humanitarian Situation in Darfur Region of Sudan.* SG/SM/9067. Available at http://www.un.org/News/Press/docs/2003/sgsm9067.doc.htm.

United Nations. 2004a. *Journal of the United Nations: Programme of Meetings and Agenda.* New York, Friday 2 April, No. 2004/64. Available at http://www.un.org/Docs/journal/En/20040402e.pdf.

United Nations. 2004b. *4978th Meeting United Nations Security Council.* New York, 25 May, S/PV.4978.

United Nations. 2004c. *Press Briefing on Humanitarian Crisis in Darfur, Sudan.* New York. Available at http://www.un.org/News/briefings/docs/2004/egelandbrf.DOC.htm.

United Nations. 2004d. *Press Statement on Darfur, Sudan, by Security Council President.* 2 April. Available at http://www.un.org/News/Press/docs/2004/sc8050.doc.htm.

United Nations. 2004e. *Report of the Security Council 1 August 2003–31 July 2004. General Assembly.* New York, 29 September 2004, A/59/2.

United Nations 2006. *5526th Meeting United Nations Security Council.* New York, 15 September, S/PV.5526.

United Nations. 2014. *7353rd Meeting United Nations Security Council.* New York, 22 December, S/PV.7353.

United Nations. 2022. UN Security Council Meetings & Outcomes Tables Dag Hammarskjöld Library. Available at https://research.un.org/en/docs/sc/quick/meetings/2022.

Verhoeven, Harry, Ricardo Soares de Oliveira, and Madhan Mohan Jaganathan. 2015. 'To Intervene in Darfur, or Not: Re-examining the R2P Debate and Its Impact'. *Global Society* 30(1): 21–37.

Vik, Cathinka. 2015. *Moral Responsibility, Statecraft and Humanitarian Intervention: The US Response to Rwanda, Darfur, and Libya*. Global Politics and the Responsibility to Protect. Abingdon and New York: Routledge.

Weissman, Fabrice. 2007. *Sudan Divided: The Challenge to Humanitarian Action*. New York: Médecins Sans Frontieres. Available at http://www.doctorswithoutborders. org/news-stories/speechopen-letter/sudan-divided-challenge-humanitarian-action.

Williams, Paul D. 2010. 'The United Kingdom'. In *The International Politics of Mass Atrocities: The Case of Darfur*, eds. Paul D. Williams and David R. Black. London and New York: Routledge.

# 5

# Inching towards Sanctions

*I am 100% certain. There was no chance of getting sanctions.*
John Danforth, former US Ambassador to the United Nations,
reflecting on his time negotiating sanctions on Darfur
in the Security Council (cited in Keane 2005)

The issue of sanctions in response to the conflict in Darfur polarized members of the Security Council more than any other aspect of the response. An arms embargo was established in Darfur for non-governmental actors in 2004 and expanded to also include the government in 2005. Eventually, the Security Council created a Sanctions Committee for Darfur and a Panel of Experts to monitor targeted sanctions, however, there was little agreement on targeted sanctions. After two years of intense negotiations over sanctions in Darfur, the Security Council issued targeted sanctions against only four individuals.[1] Permanent members had 'no common interest' in Sudan in terms of creating a robust sanctions architecture, so sanctions were weak and limited (Dörfler 2019: 150). As highlighted in Figure 2, sanctions on Darfur were one of the most divisive matters in the Security Council during this period, compared to other situations on the Council's agenda at the time. Indeed, in the period 2004 to 2006 a total of 222 resolutions were put to the Security Council, and only eleven were not unanimous. Of these eleven, six were on Darfur and four of these included provisions on sanctions, showing that sanctions were one of the most contentious issues for the Security Council in relation to Darfur.[2] This gives some perspective on how difficult it was for the Council to agree to minimal sanctions on Darfur, relative to other matters under consideration at the time. After two years of intense debate, the impact of sanctions on the ground in Darfur was negligible.

External legitimation practices are evident throughout Darfur sanctions negotiations. There is a clear pattern of behaviour from Western states with a desire to 'do something', or at least be seen to be doing something. It is common for liberal states to advocate the use of sanctions as a means of 'doing something'. As Tierney argued

---

[1] On issues with the implementation of sanctions in general, including sanctions on Darfur, see Eckert (2016) and Farrall (2009).
[2] During this period there were also five draft resolutions that were vetoed. Data compiled by the author from UN (2022).

*Inside the UN Security Council.* Jess Gifkins, Oxford University Press. © Jess Gifkins (2023).
DOI: 10.1093/oso/9780192869029.003.0006

'economic sanctions mean that something is being done; potentially gaining polit-ical capital in a domestic environment that demands action' (2005: 647). Elliott also uses this same framing: 'sanctions may also be intended to signal domestic constituencies or allies, who are looking to a country such as the United States to "do something" when outrages occur in the world' (2009–10: 87–88). The events analysed in this chapter show a consistent drive to 'do something' by Western states, however, this did not translate into marshalling effective action. Sanctions are often 'the Council's first and most enduring response' (Boulden and Charron 2009–10). The other key legitimation practice evident in this chapter is Beijing's willingness to block decisions privately—both via threatened vetoes and via block-ing decisions in the subsidiary Sanctions Committee—yet allowing decisions, in some cases the same decision, to pass rather than cast a veto in a public forum. This demonstrates China seeking legitimation, which shaped its actions. The positions taken by Western states and China in these debates demonstrate their interests in seeking—in the case of the West—the social status of legitimacy from domestic audiences—and for China—legitimacy from international audiences and peers.

The analysis in this chapter also presents a more nuanced account of state positions on Darfur. As discussed in the Introduction, academic literature often represents the US as the key advocate on Darfur within the Council and China as the key opponent. While the US was often the most vocal on Darfur, its actions in the sanctions negotiations also included pushing through resolutions which could not be implemented and reducing the number of individuals to be issued with targeted sanctions. China's foreign policy places a strong emphasis on the princi-ple of non-interference and is generally sceptical of UN sanctions; Darfur was no exception (Wenqi and Xinyu 2016; Wuthnow 2010). China opposed sanctions on Darfur by resisting the threat and imposition of sanctions in the early stages, and by limiting the individuals to be targeted with sanctions once a sanctions regime was established. There were, however, also limits to China's resistance however, whereby China was unwilling to block a decision in the Security Council that it had previously blocked within the Sanctions Committee. This shift in behaviour was associated with the shift to a more public venue, which indicates sensitivity on the part of China to the legitimation costs that come with blocking a decision in a public setting.

This chapter analyses the negotiations towards four key resolutions which threatened or applied sanctions in response to the conflict in Darfur. Subse-quently, there have been over a dozen resolutions relating to sanctions on Darfur, however, these later resolutions were simple renewals for the Sanctions Commit-tee and the Panel of Experts, with little ensuing debate or negotiation. As such, this chapter focuses on the first four decisions to include or threaten sanctions on Darfur, which spanned 2004–2006, which were highly contentious and involved lengthy negotiations. Resolution 1556 (2004) which authorized an arms embargo and threatened wider sanctions; resolution 1564 (2004) which again threatened wider sanctions; resolution 1591 (2005) which established sanctions infrastructure in the form of a Sanctions Committee and a Panel of Experts; and resolution 1672 (2006) which issued targeted sanctions against four individuals in Darfur.[3] After these four highly contentious resolutions, decisions on sanctions in Darfur became routine, with regular renewals of the mandate of the Sanctions Committee Panel of Experts.

## Early Divisions Appear: Negotiating Resolution 1556

Resolution 1556, the first sanctions resolution on Darfur, is rightly described as an abdication or abrogation of the Security Council's responsibilities in Darfur (Clough 2005: 5; Williams 2006: 178; Williams and Bellamy 2005: 32). As domestic interest in Darfur grew in Western countries there were political incentives to 'take action' and resolution 1556 gave the impression of action; it was a legally binding resolution taken under Chapter VII of the UN Charter. There were two key provisions in resolution 1556 when it passed on 30 July 2004. First, it expressed the intention to 'consider' applying further measures against Khartoum, including Article 41 of the UN Charter, which refers to sanctions, if it did not disarm the Janjaweed within a thirty-day reporting period. Second, it established an arms embargo to prevent the supply or sale of weaponry to non-government groups or individuals who were operating in Darfur; however, the arms embargo did not apply to the whole of Sudan or to the government itself, which was arming the Janjaweed militia. Resolution 1556 also failed to define who the 'Janjaweed' were or what 'disarming' them would entail, creating ambiguity over which specific actions Khartoum would be held accountable for (de Waal 2007: 1041). The substance of resolution 1556 created real challenges, with provisions that restricted the application of the arms embargo in terms of geography and actors, and lack of clarity on what compliance would look like. As the Security Council's first substantive engagement on Darfur, this resolution demonstrated a political compromise which prioritized the external legitimation practice of 'doing something' over taking meaningful action.

There were serious concerns over the efficacy of the arms embargo established in resolution 1556, as it applied to non-government groups, but not to the government of Sudan, which 'organizes, finances, directs, and supplies the Janjaweed' (Clough 2005: 5).[4] As Elliott has argued, 'When arms embargoes are revealed to be paper tigers, as so often happens, however, the United Nations' credibility is undermined' (2009–10: 88). Tierney concurs with this position, arguing that 'the

---

[3] Resolution 1679 (2006) also included the threat of sanctions. This resolution is discussed in Chapter 7, however, as it primarily focused on peacekeeping.
[4] The Security Council later expanded the arms embargo to include the government of Sudan in resolution 1591 in 2005.

best scenario is an enforced embargo, the second best scenario is no embargo, and the worst scenario is an unenforced embargo' (2005: 655). The Security Council did not create infrastructure for monitoring the arms embargo until the following year, so resolution 1556 represented the worst scenario of an unenforced arms embargo.

Fault lines emerged in negotiations towards the first draft of this resolution on sanctions in Darfur. In June 2004 the US circulated a draft which called for an arms embargo and travel ban on the Janjaweed militia, with the possibility of later sanctions on the government (AP 2004b). This early draft was criticized by China, Russia, Pakistan, Algeria, and Brazil for being too strong (Reuters 2004). Conversely, a group of Western states—the UK, France, and Germany—argued that the proposal did not go far enough and that the embargo should apply to all of Sudan, rather than only Darfur (Reuters 2004). It was evident right from these early negotiations that it would be quite difficult for the Security Council to find a meaningful compromise.

Negotiations stalled for a month after the first draft, during which an agreement was reached between the United Nations and the government of Sudan, known as the Joint Communiqué. In this agreement, Khartoum committed to 'immediately start to disarm the Janjaweed militia and other armed outlaw groups', and some Security Council members argued that UN sanctions could be used to enforce this commitment (United Nations 2004c: 3). The Joint Communiqué then formed the basis for the next round of negotiations in the Security Council. Over the space of a week in late July, the US released a series of drafts which strengthened its original proposal. These drafts threatened unspecified sanctions against the government of Sudan if it did not disarm the Janjaweed militia, as agreed to under the Joint Communiqué, and included a thirty-day reporting cycle (AP 2004a; Lynch 2004). The key points of contention between Council members at this stage were the inclusion of sanctions against Khartoum and the timing of the resolution. China, Russia, and Pakistan opposed the use of sanctions and wanted Khartoum to have more time to meet its commitments (Gamel 2004). Far from resolving divisions, this stronger draft from the US increased the extent of the division between Council members. By 29 July, the US draft was opposed by seven members of the Council—China, Russia, Algeria, Angola, Brazil Pakistan, and the Philippines (BBC 2004). This group not only included permanent members with the capacity to veto, but also created the possibility that the draft would fall short of the legal majority of nine, if those opposed were not persuaded to vote affirmatively. Division over what strategy to use meant that it was increasingly difficult for drafters to find an acceptable solution.

In the absence of agreement between Security Council members, the impasse was breeched with a 'text-based compromise', whereby the word 'sanctions' was removed from the draft. This compromise was recommended to the US by other Security Council members. The Chinese Deputy Ambassador Zhang Yishan said that 'quite a number' of Security Council members had suggested removing the word 'sanctions' from the draft, including Beijing (Leopold 2004). The word 'sanctions' was removed and replaced with 'measures as provided for in Article 41 of the Charter of the United Nations' (United Nations 2004e: 3). Article 41 refers to curtailing economic interactions, communication and diplomatic relations (United Nations 1945). As US Ambassador John Danforth explained, Article 41 is 'UN speak' for sanctions (2004b). Indeed, the UN Charter doesn't use the word sanctions (Tzanakopoulos 2014). This text-based compromise enabled the resolution to pass but created serious challenges for implementation.

The text-based compromise did not resolve the underlying issue: that China and other states were unwilling to impose broader economic sanctions on Sudan. US Ambassador Danforth made it clear even without the word 'sanctions' the threat of sanctions was still there from the perspective of Washington. When asked about the implications of removing the word sanctions from the draft, Ambassador Danforth said 'Clearly, if there is no compliance by the government of Sudan ... then the Security Council would have to act, and those actions would have to include sanctions' (Danforth 2004a). He continued, 'So if you wanted to use the word banana, so long as it's clear it equals sanctions' (Danforth 2004a). Downgrading this language suited Khartoum well. After the vote, Sudan's Foreign Minister Mustafa Ismail bluntly described what its strategy had been during these negotiations: 'either to block its adoption altogether or to strive, in cooperation with our friends, to remove from it such references as genocide, ethnic cleansing and other extreme points and apparently this is what we have so far succeeded in achieving just hours before the vote' (Daily Star 2004). Sudan maintained that it had succeeded in making the text ineffective, while the US maintained that sanctions were still on the table. The text-based compromise allowed states to maintain their own interpretations of what the resolution meant.

Statements made within the Security Council after the vote on resolution 1556 confirm that the issue of sanctions remained unresolved. States held different perspectives, not only on whether threatening sanctions was appropriate, but whether the resolution even included that threat. This was most evident in statements made by the US and Russia. The US stated explicitly that 'The resolution anticipates sanctions against the Government of Sudan if the regular monthly reporting cycle reveals a lack of compliance' (United Nations 2004a: 4). Whereas Russia, alluding to sanctions, said 'of fundamental importance is the fact that the resolution does not foresee possible further Security Council action with regard to Darfur' (United Nations 2004a: 7). Drafters found language that Council members could accept, but there were divergent interpretations on what the resolution meant. Statements by the UK, France, Germany, and Spain followed a similar line to the US, arguing that sanctions would be unavoidable if Khartoum did not comply (United Nations 2004a: 5–9). Statements from China and Pakistan disputed the appropriateness of threatening sanctions against the government of Sudan at that point in time, but unlike Russia, they did not dispute that the threat of sanctions was included in the resolution (United Nations 2004a: 2–10). As aptly described by Bellamy, the resolution was 'Janus-faced' and meant different things to different participants (2005: 43). The policy was far from clear in this case, and as Wood has argued, 'It is, of course, only possible to use clear language when the policy is clear' (1998: 82). The text-based compromise—driven by the desire to 'do something', even if it could not be implemented—enabled the resolution to pass but masked real differences in perspective on sanctions against Sudan.

In this case the resolution threatened sanctions if the government of Sudan had not fulfilled its commitment to disarm the Janjaweed militia within thirty days, but unsurprisingly, a month later when it was found that Khartoum had not met this target sanctions were not implemented. In fact, a diplomat who was involved in the negotiations towards resolution 1556 said that 'few believed it would be different in thirty days' (Author Interview 2011b). By prioritizing unity over clarity, drafters passed a resolution with a 'hollow' threat, i.e., a threat to the credibility of the Security Council. The Security Council relies on the credibility and legitimacy of its decisions for implementation, so this represented a problem for the institution (Hurd 2002, 2007; Voeten 2005). In passing this resolution, the Security Council demonstrated that there was a lack of political will to implement sanctions against Sudan at this time, beyond the limited arms embargo, as the negotiations towards the next resolution made painfully clear.

By the time the Security Council passed resolution 1556 it had demonstrated to the outside world—and more importantly to Khartoum—that it was deeply divided on the question of sanctions. This was demonstrated not only via the removal of the word 'sanctions' from the resolution but also in statements made to the media by both those who supported and opposed sanctions. A British civil servant, which supported sanctions, reported: 'There is a range of views in the Security Council, but the natural centre of gravity is not the imposition of heavy duty sanctions on Sudan' (Xinhua 2004). Pakistan, which opposed sanctions, stated more bluntly, 'Frankly, at the moment, there is not the political will' (Kirkup 2004). By prioritizing 'doing something'—or more specifically, being seen to be doing something—over taking implementable action, the Security Council demonstrated to Khartoum that it did not pose a serious threat to its plans in Darfur. In fact, there is even a suggestion that the rate of deaths in Darfur increased again after resolution 1556 passed (Grünfeld and Vermeulen 2009: 230). An interviewee from a think tank based in New York said that 'the Security Council had to act for domestic purposes' of individual member states (Author Interview 2011a). By the time resolution 1556 passed, sanctions had already been discussed multiple times by the US Congress, and the key driver of US interest in this case was domestic civil society groups (Stedjan and Thomas-Jensen 2010: 158 and 162). There was a strong legitimation incentive to be seen to be 'doing something', but insufficient unity or diplomacy to craft a united response.

## Divided on Sanctions: Negotiating Resolution 1564

Given the negotiations towards resolution 1556, and statements made thereafter, it was not surprising that negotiations towards the follow-up resolution were also fraught. Resolution 1556 had asked the UN Secretary-General to report each month on Khartoum's compliance with disarming the Janjaweed militia and bringing participants to justice. The first monthly report made a clear assessment that 'Attacks against civilians are continuing and the vast majority of armed militias have not been disarmed' (United Nations 2004d: 14). Softening this statement however, the report stated (not unreasonably) that there was agreement between the UN and its partners 'that not all commitments made by the Government could be fully implemented within 30 days' (United Nations 2004d: 2). This speaks to the difficult situation the Security Council had created for itself in resolution 1556 by giving Khartoum only thirty days to disarm the Janjaweed militia, and not defining how it would measure this. The report described positive cooperation between the UN and Khartoum in the planning stage, but the key question was implementation, and on this there had been little progress (United Nations 2004d: 4). This assessment presented the Security Council with a bind of its own making; Khartoum had only made minimal progress towards the demands made in resolution 1556, yet there was not sufficient political will within the Council to implement sanctions for non-compliance. Responding to this bind, drafters again negotiated a resolution by finding language that Council members could accept, rather than reaching a substantive agreement. Again, in these negotiations, drafters prioritized 'doing something' over taking action that could be implemented.

As with the previous resolution, the initial drafts had strong provisions which were conceded by the US as negotiations progressed. For example, the first draft 'demands' that the government cease military flights over Darfur while the second draft 'calls on the government of Sudan to cease all military flights' (Lederer 2004), and the final resolution 'urges the Government of Sudan to refrain from conducting military flights' (United Nations 2004f: 3). It may appear minor to downgrade the language in a resolution from 'demands', to 'calls on', to 'urges', however, this language has legal implications which are well known to those involved in the negotiations. The specific wording of a resolution determines whether the decision is legally binding. The think tank Security Council Report explained this general principle: 'it can be clearly established that by using "urges" and "invites" as opposed to "decides" the paragraph is intended to be exhortatory and not binding' (Security Council Report 2008: 9). Thus, during the negotiations the provision on military flights went from a legally binding 'demand' to a non-binding 'request'. Like the previous negotiations, drafters prioritized 'doing something' over creating a resolution which could be meaningfully enforced.

The language on sanctions was also downgraded during negotiations. The first draft said that the Council 'will take further actions' if Khartoum failed to comply with the resolution, suggesting automaticity, while the second draft used the milder phrase 'shall consider taking additional measures', which was retained in the final version (Lederer 2004). This change moved from an expectation of sanctions in the event of non-compliance, to a possibility that sanctions would be applied at an unspecified point in the future. This language was changed at the insistence of Beijing (Holslag 2008: 82; Taylor 2010: 180). Using its institutional power, China threatened to veto the first draft of what became resolution 1564 (El-Tablawy 2004; Lederer 2004). Although Pakistan was vocal in its opposition to the early draft, it was Beijing's opposition, with the capacity to veto, that meant this was taken seriously (Traub 2010: 11). China did not exercise a veto, which would go on the public record, but threatened to vote against the draft. A negative vote from China would have blocked the draft, so drafters took the threat seriously.[5] By using this institutional power, China was able to strip any automaticity of sanctions and to change the provision on military flights to a request, without incurring the reputational costs and publicity of a negative vote.

By this stage, US rhetoric on Darfur was strong. US Secretary of State Colin Powell stated in the US Senate, 'we concluded, I concluded, that genocide has been committed in Darfur, and that the Government of Sudan and the Janjaweed bear responsibility and that genocide may still be occurring' (Committee on Foreign Relations 2004). However, Powell continued: 'some seem to have been waiting for this determination of genocide to take action. In fact, however, no new action is dictated by this determination'. The decision to describe the situation as genocide was made by Powell alone, and represented the first time the US executive branch had declared a situation to be genocide while it was ongoing (Hamilton 2011b).[6] This was a strong statement from Powell; however, it did not translate into action.

Given the genocide determination by the US executive, Bellamy argues that the remarkable thing is 'not so much that the resolution was toned down to secure consensus, but that the United States especially chose not to argue along the lines of Romania and the Philippines', which made statements on international responsibilities linked to the then nascent 'responsibility to protect' idea (2005: 47).[7] A US diplomat involved in the negotiations for resolution 1564 said 'We never went beyond rhetoric in offloading this to the UN' (Traub 2010: 11). When asked about the US determination of genocide in Darfur, Danforth commented, 'I just thought that this was something that was said for internal consumption within the United States' (Keane 2005). With advocacy from civil society, activist members in

---

[5] At the time of writing, China had not exercised a veto on its own since the 1990s. Vetoes cast by Beijing since this time had been alongside Russia.

[6] For further discussion on the 'genocide debate' see (Brunk 2008; Glanville 2009; Heinze 2007; Gifkins 2016).

[7] The International Commission on Intervention and State Sovereignty released its report titled 'The Responsibility to Protect' in 2001, however, the General Assembly did not agree to a version of R2P until the 2005 World Summit outcome document. This resolution was negotiated between these two events.

Congress, and a president who was elected after making promises to take action on the southern region of Sudan (now South Sudan) there were incentives for the US to take action, or at least, look like it was taking action (Stedjan and Thomas-Jensen 2010). Specifically, 'The movement created a small political cost and the administration responded with rhetoric and some diplomacy, but US actions were never sufficient to generate an effective international response' (Stedjan and Thomas-Jensen 2010: 161). The US made strong statements which looked like it was 'doing something', but conceded to China on following through on sanctions.

There were two key aspects of resolution 1564 when it passed on 18 September 2004, with abstentions from China, Russia, Pakistan, and Algeria (United Nations 2004f).[8] The most significant provision was that it authorized the establishment of an International Commission of Inquiry on Darfur (ICID) to investigate human rights violations, and specifically, whether 'acts of genocide' had occurred. In conducting this investigation, the commission was also tasked with identifying perpetrators of human rights abuses, with a view to holding individuals to account. This commission, and its findings, went on to shape Council negotiations a year later in 2005 towards the referral of Darfur to the International Criminal Court (ICC).[9] The other key aspect of resolution 1564 was that it stated that 'the Government of Sudan has not fully met its obligations noted in resolution 1556', and that it would 'consider taking additional measures' if Khartoum did not meet its obligations under resolutions 1556 and 1564. This resolution went further than resolution 1556 in explaining the type of sanctions that were being considered; targeted sanctions against individuals and against the oil industry (United Nations 2004b: 4). However, given the hollow threat in the earlier resolution, this further signalled to Khartoum that there was no political will to issue targeted sanctions at this time, despite clear evidence that Khartoum had not met its obligations. The threat of a veto from China—using its institutional power—effectively took further sanctions off the table at this stage of the negotiations.

The inclusion of the ICID in the resolution was driven by the US and Powell's determination on genocide (Grünfeld, Vermeulen and Krommendijk 2014: 159). There is a discontinuity here, however, between determining that the situation constituted genocide and then mandating an inquiry to determine, among other things, whether acts of genocide had occurred. Creating additional investigations was a weak response, which served to give the appearance that action was being taken: looking like they were 'doing something' (Guéhenno 2015: 165–166; Totten

---

[8] Resolution 1564 also contained the phrase 'responsibility to protect', linked to the then emerging principle outlined in the 2001 report by the International Commission on Intervention and State Sovereignty (ICISS). The Philippines also echoed the ICISS report in its statement after the vote. This did not represent formal use of R2P language, as this predated the 2005 World Summit and there was no agreement on the meaning of R2P at this time, but it did reflect growing awareness of the concept for individual Security Council members. For further discussion see Gifkins (2016: 5–6).

[9] The findings of this commission and its effects are discussed in detail in Chapter 6.

2010: 199). Perhaps the most generous interpretation is that the role of the ICID was not so much to determine the facts on the ground but to build pressure and buy time (Guéhenno 2015: 166). The ICID investigation delayed further discussion on violations of international law until 2005.

These negotiations were again driven by the desire on the part of Western states to 'do something' even if the actions themselves were minimal. Considering the later referral of the situation in Darfur to the ICC, this resolution could, at best, be seen as a stepping stone towards more substantive action; however, on sanctions, the response continued to undermine the credibility of the Security Council. China's exercise of its institutional power—in the form of a threatened veto—meant that the question of sanctions was off the table at this stage, despite clear evidence that Khartoum had not met the requirements established under resolution 1556. By 'doing something'—prioritizing the appearance of action over substantive action—the Security Council demonstrated to Khartoum that it did not represent a threat in the form of sanctions.

## Establishing Sanctions Infrastructure: Negotiating Resolution 1591

The next push on sanctions, again led by the US, began at the start of 2005. This time the timing was right; by early 2005, the North–South peace deal was in place, giving the Security Council capacity to pivot back to Darfur, and the findings of the ICID had been released (Dörfler 2019: 153). In anticipation of the ICID report, the US began to discuss plans for a new resolution with other permanent members which crystallized around three key areas: increased sanctions in relation to Darfur, deployment of peacekeepers to southern Sudan to monitor the Comprehensive Peace Agreement (CPA), and a tribunal to hold individuals accountable for crimes committed in Darfur (Mohammed 2005). The ICID report was completed 25 January 2005 and released to the Security Council the following week. While it did not find evidence of a government policy of genocide, it was clear that 'the crimes against humanity and war crimes that have been committed in Darfur may be no less serious and heinous than genocide' and recommended that the situation in Darfur be referred to the International Criminal Court (United Nations 2005b: 3–5). It also contained a document with a (sealed) list of fifty-one names of people suspected of crimes under international humanitarian law and international human rights law (United Nations 2005b: 161). The release of this report gave fresh impetus to debates both on targeted sanctions and on international criminal accountability.[10]

---

[10] For extended discussion on the decision to refer the situation in Darfur to the International Criminal Court, see Chapter 6.

During negotiations over an initial triptych resolution—sanctions, peacekeeping, and accountability—both Russia and China spoke out about their opposition to the use of economic sanctions (Interfax 2005; Xinhua 2005). The draft circulated by the US in mid-February included provisions for targeted sanctions for individuals implicated in violence, authorized a Sanctions Committee, broadened the arms embargo to include the government of Sudan, and included the threat of a future oil embargo, alongside a peacekeeping operation in southern Sudan and a determination to bring perpetrators to justice by 'internationally accepted means' (Reuters 2005). The previous arms embargo established under resolution 1556 applied to all non-governmental actors, however, the US and UK pushed to widen the arms embargo to cover all of Sudan, rather than only Darfur (Turner 2005). There were real divisions on the question of the arms embargo, as China and Russia did not want the arms embargo expanded to include the government of Sudan (Interfax 2005; Lederer 2005). The question of an oil embargo for Sudan was also particularly contentious for both Russia and China and was conceded during negotiations and replaced with coded language on sanctions: 'will consider further measures as provided for in Article 41 of the Charter of the United Nations' (Leopold 2005a). After leading negotiations on the three-part resolution for months and getting nowhere, the US reached a point of frustration and split the draft into three separate drafts on peacekeeping for the south of Sudan: the least contentious aspect (which became resolution 1590), sanctions for Darfur (which became resolution 1591), and accountability in Darfur (which became resolution 1593) (Hoge 2005).

There is less information available on the process of negotiating resolution 1591 than the previous two resolutions—for two reasons. First, the most contentious aspect of the triptych was international accountability, rather than sanctions or peacekeeping in southern Sudan, so there was more attention at the time on the question of accountability. In that sense, splitting the initial draft into three was crucial to move the conversation forward. Second, it was reported on the 24 March that the draft had been split into three drafts and resolution 1591 passed only five days later (Hoge 2005). This meant that the final stage of negotiations was quite quick, with fewer statements or drafts in between. What the evidence shows, however, is that there were two changes that facilitated the passage of resolution 1591, splitting the draft into three and removing the threat of an oil embargo on Sudan which Russia and China had opposed during informal negotiations.

A comparison between the sanctions section of a (leaked) US draft from 14 February with the final text of resolution 1591 reveals that the only significant change regarding sanctions was the removal of text concerning the oil industry. The mid-February draft stated that the Security Council 'will consider further measures as provided for in Article 41 of the Charter, including through actions relating to Sudan's petroleum sector', which was edited in the final version to read: 'will consider further measures as provided for in Article 41 of the Charter of the United Nations' (Global Policy Forum 2005; United Nations 2005c). This downgrading of language mirrored the negotiations towards the previous resolution, where a threat of sanctions against the oil industry was also removed. A provision on the need for increased human rights monitors in Darfur was also removed (Global Policy Forum 2005). The authorized version extended the arms embargo—not to the whole of Sudan—which China and Russia had opposed, but to all 'the parties to the N'djamena Ceasefire Agreement and any other belligerents in the states of North Darfur, South Darfur and West Darfur' (United Nations 2005c: 5). This meant that the arms embargo applied to the government of Sudan, as a party to the N'djamena Ceasefire Agreement; however, it still only applied to Darfur rather than Sudan as a whole. This meant that the government could still import arms if it did not transfer them to Darfur. Unsurprisingly, this has proven almost impossible to monitor (Dörfler 2019: 155).

The primary achievement of resolution 1591, however, was that it established a Sanctions Committee and Panel of Experts to monitor the implementation of resolution 1556 and resolution 1591 (United Nations 2005c). The creation of a Sanctions Committee and Panel of Experts gave heft to the previous decisions on the arms embargo and paved the way towards targeted sanctions against individuals. This division of labour meant that a four-person Panel of Experts would conduct research and make recommendations for individuals to be subject to sanctions, and the Sanctions Committee which would serve as a 'filter' between the Panel and the Security Council (Farrall 2009: 207). The Security Council asked the subsidiary Sanctions Committee to designate individuals to be subject to sanctions and to report to the Council.

Resolution 1591 was stronger than previous resolutions on sanctions in three ways. First, the resolution was no longer tiptoeing around the issue of sanctions but authorized the infrastructure for implementation in the form of a Sanctions Committee and a Panel of Experts. Under the new provisions, the Security Council could issue sanctions against individuals in form of asset freezes and travel bans. Second, unlike previous resolutions which contained a threat of future sanctions, resolution 1591 had automaticity whereby measures would come into effect in thirty days *unless* the Security Council determined that parties to the conflict were complying with their commitments, which circumvented the need for additional negotiations (United Nations 2005c: 5). Third, the resolution expanded the arms embargo from non-governmental actors to all parties to the N'djamena Ceasefire Agreement, which included the government of Sudan. This was an important shift, as the government was known to be arming the Janjaweed militia, so an arms embargo that did not include the government was not likely to be effective. There remained issues with the implementation of all these aspects—from delays in appointing the Panel of Experts to violations of the arms embargo[11]—but the resolution at least represented a more proactive step from the Security Council.

---

[11] On difficulties in enforcing the arms embargo see Vine (2007). On delays in establishing the Panel of Experts and Sanctions Committee see Farrall (2009).

Statements made in the Security Council after the vote again highlighted some of the sticking points from the negotiations. China, Russia, and Algeria abstained from the vote and all three said that their requested amendments were not incorporated and stressed the need for the Security Council to play a constructive role. They expressed concerns about the timing of the resolution and that it might jeopardize the north/south peace process. The statements from China and Russia are most instructive, as each of them had the capacity to block the decision but chose not to. Like the previous resolution on Darfur sanctions, China expressed 'serious reservations' about the resolution (United Nations 2005a: 4). Similarly, Russia stated 'The imposition of sanctions against that Government is hardly likely to create a constructive atmosphere for its efforts' (United Nations 2005a: 4). Despite Beijing and Moscow's ongoing and serious concerns at the time of the vote, they abstained, and by doing so enabled the resolution to pass.

Why did China, which had been the primary opponent of a robust sanctions regime in Darfur, enable this resolution to pass, when it retained serious concerns about the decision? The most common, and not wholly unreasonable, answer to this question is the rationalist position that China was able to achieve much of what it wanted in the form of concessions during negotiations (Dörfler 2019; Holslag 2008; Wuthnow 2010). The threat of a future oil embargo was removed, the arms embargo was extended to the government, but only in Darfur, which left Khartoum plenty of room to manoeuvre, and by moving decisions on targeted sanctions to a Sanctions Committee, China held considerable leverage in the process. Indeed, Dörfler argues that 'Transferring political conflict over Darfur into the sanctions committee was an ideal way of deflecting international pressure away from the Council' (2019: 155). Of all the negotiations analysed in this chapter, resolution 1591 presents the strongest challenge to my argument on how legitimation practices shape negotiations. It is clear that the rationalist position that China, and Russia, achieved what they wanted in the negotiations is a part of the explanation for how this resolution was achieved. There are two other constructivist arguments on China's acquiescence on Darfur. Wuthnow argues that China was sensitive to reputational concerns, as it attempted to establish itself as a 'responsible' power in Africa (2010: 71). Holslag's position is that China has undergone a significant process of socialization and learning in the Council over its decades of membership (2008: 74). In addition to these existing constructivist arguments, there is an additional legitimacy-based explanation for this resolution, even if it is less strong than for other negotiations.

The two aspects of my framework that are relevant to the negotiations towards resolution 1591 are the consistency of arguments, and the desire to 'do something'. China's statement after the vote also cites other legitimacy-based practices such as the importance of supporting relevant regional groups (United Nations 2005a: 5). Indeed, regions often lead with applying sanctions followed by the UN (Charron and Portella 2015). For the US, the desire to 'do something', or at least be seen to be doing something, was strong on Darfur. By that stage, it had declared that genocide had occurred in Darfur and may be ongoing. It had deferred hard decisions on accountability in resolution 1564 by advocating the ICID, but once the ICID had determined that war crimes and crimes against humanity were occurring in Darfur, it was then very difficult to not take action in response. By this stage, momentum had been built via regular briefings and meetings. There is a sense in which the Security Council undergoes a process of ratcheting up through a fairly standard sequence of tools at its disposal (Gehring and Dörfler 2019). China was clear at the time of the vote that it still maintained 'serious reservations', yet it did not use the power that it had to prevent the resolution from passing.

In its statement after the vote, China did not dispute the magnitude of crimes being committed in Darfur but expressed general concerns around sanctions and the negative impact they could have on the peace process (United Nations 2005a: 5). Russia expressed similar concerns about the current situation on the ground in Darfur but raised questions about the timing of sanctions (United Nations 2005a: 4). In light of determinations that serious international crimes had been committed in Darfur, both from the US and the ICID, and ratcheted pressure from the Security Council, it would have been difficult for China or Russia to veto this draft while maintaining the position that they were concerned about violence in Darfur. Momentum had been built on this issue in the Security Council for a year by this stage, with regular briefings and meetings and slowly ratcheting up through the Council's tools. This, in addition to the concessions achieved by China and Russia during negotiations, made it difficult for them to oppose the decision to create sanctions infrastructure.

## Designating Individuals for Targeted Sanctions: Negotiating Resolution 1672

Following the passage of resolution 1591—which authorized targeted sanctions but did not determine who they would be applied to—the process still involved lengthy delays and obstruction from China, Russia, Qatar, and the US. The Panel of Experts eventually recommended sanctions against seventeen individuals; however, the Security Council could only agree on sanctions against four of these people. The year-long process towards resolution 1672 reveals interesting behaviours in the Council, whereby China and Russia were willing to block sanctions even against four individuals in the Sanctions Committee but were unwilling to exercise a public veto on the same decision once it was put to a vote in the Security Council, thereby allowing the decision to pass. The key change here was that Russia backed away from issuing a veto and China reported privately to US diplomats that it would be 'unthinkable' for China to block the resolution against the wishes of fourteen members of the Security Council (US Embassy Khartoum 2006). Although individual targeted sanctions were authorized, it is worth remembering that the content of the resolution was largely symbolic as sanctions were only applied to four mid-ranked officials (Dörfler 2019: 161–163).

The process of reaching agreement on resolution 1672 highlights two key aspects of behaviour in the Security Council. First, China and Russia were willing to block privately what they were not willing to block publicly. This shows that external legitimation practices have more influence in public negotiations than in private negotiations. China and Russia did not want sanctions applied in this case, however, there was greater consideration of 'appearances' once negotiations were moved to a public arena. It also suggests that the Security Council and its subsidiary bodies have discrete social environments (Gehring, Dorsch and Dörfler 2019). Second, once China was isolated, it was not willing to exercise a public veto. When China has vetoed decisions, it has typically been in retaliation for perceived threats to its 'One China' policy (Wenqi and Xinyu 2016: 86 and 90–91). Indeed, at the time they were negotiating resolution 1672 in 2006, the last veto China had cast was in 1999.[12] This shows that while China was willing to use its informal veto during negotiations in the Sanctions Committee, it was less willing to cast a public veto—especially an isolated veto—as there are greater political and reputational costs for this. Seeking the social status of legitimacy led China to allow the resolution to pass once negotiations moved to the public Security Council rather than the subsidiary Sanctions Committee.

Delays and obstructions towards resolution 1672 were extensive. The Panel of Experts was authorized in March 2005, yet it wasn't until July 2005 that a panel was even created after some candidates turned down the role and others were blocked by China (Leopold 2005b). The four-person Panel of Experts issued its report to the Sanctions Committee in December 2005, but the report was blocked from transmission to the Security Council by China and Qatar (Arieff 2006a, 2006b). Following these delays, it was not until the end of February 2006 that the report was scheduled for its first discussion in the Security Council (All Africa Global Media 2006). This was almost a full twelve months after the Panel of Experts was authorized, after considerable obstruction from China.

After the panel's report was transmitted to the Security Council, it took a further six months for the Council to act on the report, following further delays by China, Russia, Qatar, and the US. The onus is often placed on China and Russia for whittling down the list of individuals to be sanctioned (for example see Holslag 2008: 81; Taylor 2010: 180), however, there is evidence that the US was directly responsible for rejecting at least four of the seventeen candidates for sanctions. The Panel of Experts' final report recommended that targeted sanctions be applied to seventeen individuals, with an additional five people to be considered for future sanctions (All Africa Global Media 2006). After negotiations between

---

[12] China vetoed draft resolution on the Republic of Macedonia on 25 February 1999.

Council members, the UK put forward a compromise list of eight of the seventeen, however, at this stage the US was only willing to apply sanctions against two individuals (Dinmore, England and Turner 2006). A key difficulty for the US was their intelligence relationship with Khartoum, including a direct relationship with Sudan's intelligence chief Sala Abdallah Gosh, who was on the panel's list for sanctions (Leopold 2006a). The US eventually agreed to submit four individuals for sanctions—two government officials and two rebels—and the list was given to the Sanctions Committee, where it was put under forty-eight-hour silence procedures (AFP 2006). The 'silence procedure' means that if no member objected, it would be authorized at the end of the two-day window. Sanctions Committees require unity in decision-making, however, which gives every member of the Committee a veto. Each member of the Security Council is represented in each Sanctions Committee, meaning that the required standard of unity is higher than in the Security Council itself. In this case, both China and Russia objected after the four names were put under silence, which blocked authorization of sanctions against those individuals (Lederer 2006). With the Sanctions Committee blocked, advocates of sanctions turned back to the Security Council.

To circumvent the Sanctions Committee, the US drafted a Security Council resolution to apply sanctions against the four individuals it was willing to accept. Days before the draft was put to a vote, it remained a real possibility that China and Russia would veto (Leopold 2006b). This came down to the wire, and both countries had already demonstrated their willingness to block the same initiative in the Sanctions Committee. At the time of the vote, China, Russia, and Qatar still had strong concerns about the resolution regarding process, suitability of sanctions as a tool, and the timing of the resolution alongside the Abuja peace talks (United Nations 2006). Yet these three states chose to abstain from the vote rather than vote against the resolution. As Johnstone explains (2011: 61) 'a Council member says in a private meeting, "we will push this to a vote", then other members must consider whether their positions and explanations will pass muster with the outside world'. Ester concurs that 'the norm against expression of self-interest will be stronger in public settings than if the debates are conducted behind closed doors' (1995: 249). The public nature of positions taken in the Security Council changed Beijing's willingness to block this action.

A Chinese diplomat explained to the US' UN delegation why its strategy had changed since it blocked this move in the Sanctions Committee; that 'once Russia and Qatar decided to abstain, Ambassador Wang (Guangya) [China's UN Ambassador] decided it was better to abstain than veto' (US Embassy Khartoum 2006). The Chinese diplomat reported that blocking the Security Council vote against the wishes of the rest of the Council was 'unthinkable' and the US diplomat writing the cable commented that 'Beijing may be willing to join others in backing Khartoum, but appears reluctant to make a stand on its own' (US Embassy Khartoum 2006). The use of the word 'unthinkable' is interesting here. The Chinese diplomat was speaking to a US diplomat who had an interest in Beijing not opposing the draft—so they may have been telling the US what it wanted to hear—but it also suggests that China recognizes that the act of individually blocking decisions within the Security Council is inappropriate or goes against usual practices. This interpretation of Beijing's language is also supported by China's voting record; at the time of writing, in the twenty-first century, China had only vetoed alongside Russia and had not cast any 'lone' vetoes.[13] This suggests that China is reluctant to block decisions when there is otherwise sufficient support for the decision to pass. As discussed in Chapter 2, there is a drive towards unity within Security Council negotiations. The inverse of this is that there is reluctance to veto otherwise consensus resolutions. China sought the legitimation benefits of enabling the resolution to pass and sought to avoid the legitimacy costs of casting a lone veto in a public forum.

This sequence of events demonstrates that the public Security Council exerts different pressures on states than the less public subsidiary Sanctions Committee. While China and Russia were willing to block this matter in the Sanctions Committee, which has little public scrutiny, their calculations were different within the Security Council itself. For Russia, the costs of vetoing were clearly different, as it decided to allow the decision, and for China the prospect of casting an isolated veto was sufficiently costly to lead it to abstain. The 'costs' here are social and reputational: they are about the legitimation of China as a responsible actor. China did not want to stand against fourteen members of the Security Council, and in this case, there were civilizing effects of publicity, given that vetoes cast in the Council are headline news, while vetoes cast in a Sanctions Committee receive little scrutiny in the media (Elster 1995, 1998). All fifteen Security Council members are represented in the Sanctions Committee, and in this case the same decision was put to both forums. It is striking that the boundaries of legitimate behaviour were different for China within the Security Council itself, and that it was 'unthinkable' for China to stand against the other members of the Security Council.

## The Security Council and Sanctions

Throughout the negotiations towards these four resolutions, Western states demonstrated again and again a desire to 'do something'—or at least be seen to be doing something—but little was achieved. There was growing domestic pressure, especially in the US, for action to be taken in response to violence in Darfur (Hamilton 2011a). The steady escalation of resolutions on Darfur, led by Western states, gave the appearance of action for domestic audiences, even if there was little substantive change. An interview with John Danforth, who was US Ambassador

---

[13] Data compiled by the author from UN (2022).

to the UN when resolutions 1556 and 1564 were negotiated, is quite illuminating. He said it was clear that 'we would consider measures, not that we would impose measures, but there was no doubt in my mind that we were not going to impose sanctions because at least the Chinese would have vetoed the sanctions. I am 100% certain. There was no chance of getting sanctions' (Keane 2005). The US led the negotiations towards these resolutions, so it is telling that, despite the threat of sanctions in these resolutions, the US Ambassador at the time believed it was not possible that sanctions would be implemented. Debates on sanctions in Darfur occupied a significant proportion of the Security Council's time during 2004–2006. However, two years of debates on sanctions only achieved a limited arms embargo on Darfur, not the whole of Sudan, and targeted sanctions against four individuals. Targeted sanctions have not been implemented, and the arms embargo has been circumvented (Dörfler 2019: 155 and 161–163). The real-world implications of these debates were negligible.

There was, however, a domestic legitimation value for these negotiations. US rhetoric on Darfur was strong, after the initial period where the US resisted any discussion of Darfur in the Security Council, including a determination of genocide. Statements made by the US, however, were more closely driven by domestic politics than they were motivated by a determination to generate an effective multilateral response (Seymour 2014). Similarly, the UK issued strong rhetoric on Darfur, but it was a low priority for the UK compared to other issues at the time, particularly the Iraq War (Williams 2010). Rhetoric was used as an external legitimation practice to bolster the impression of action on Darfur with domestic constituents to give the appearance of 'doing something'.

China, like Western states, also demonstrated sensitivity to reputation and legitimation. Debates on sanctions in Darfur illustrate the institutional power held by permanent members. During these debates, China did not issue any formal veto votes in the Security Council. It issued threats of vetoes in the Security Council and blocked a decision in the subsidiary Sanctions Committee, alongside Russia and Qatar. While these actions have the effect of preventing a decision, they do not receive the level of attention and condemnation that Security Council vetoes receive. By blocking a decision in the Sanctions Committee and threatening to veto in the Security Council, China used its institutional power to create the outcome it wanted without the negative scrutiny of a veto. Interestingly, China then showed sensitivity to negative international attention by reversing its decision and allowing targeted sanctions to be imposed once it was isolated and once the venue was changed to the Security Council itself. That a Chinese diplomat described it as 'unthinkable' for China to veto a decision that would otherwise pass is telling of China's sensitivity to this kind of criticism; indeed, at the time of writing, China had not cast a lone veto for decades. In the case of sanctions on Darfur, China has demonstrated willingness to block decisions when they were informal and off the record, but lack of willingness to cast isolated negative votes which would carry reputational costs.

## Conclusion

The concept of external legitimation practices can be used to explain key positions in sanctions debates on Darfur, particularly Western advocacy and Chinese acquiescence. The US and UK had limited interests in Darfur as a foreign policy priority, but they did have interests in appeasing the growing domestic advocacy and civil society groups in Darfur by taking some form of action. Throughout Security Council negotiations on sanctions, there was a strong impulse from Western states to 'do something', or at least be seen to be doing something, which had legitimation benefits for them as actors, but made little difference in Darfur. Two years of intense debates over sanctions in Darfur—spanning four resolutions—achieved close to nothing on the ground in Darfur. The appearance of action was maintained, however, via statements, the US genocide determination, and regular Security Council activity. China's position, too, can be explained via legitimation practices. It sought to avoid the reputational costs of publicly blocking decisions, so it acted in the shadows using the institutional power of informal vetoes, and by blocking a decision in the less public Sanctions Committee. When there was a potential cost for China in blocking action—once activity moved to the Security Council itself—China succumbed to international pressure and enabled the resolution to pass. As part of the external legitimation practice of making consistent arguments, it was difficult for China to maintain the position that it was concerned about violence in Darfur if it had publicly blocked minimal targeted sanctions on Darfur. These are the key actions to explain in Security Council negotiations towards sanctions in Darfur—Western advocacy with little effect, and Chinese resistance which was eventually overcome to enable sanctions infrastructure at least. The frame of external legitimation practices shows how and why these positions were taken and how ineffective action was enabled.

While Security Council debates on Darfur are typically explained via rationalist and material arguments, this chapter shows that social and relational legitimation practices can instead be used to explain positions and outcomes in negotiations over sanctions. Western states were sensitive to perceptions of their actions domestically and China was sensitive to reputational costs internationally. The influence of legitimation practices on decision-making does not necessarily mean that decisions are better or more effective. As the sanctions negotiations show, some negotiations resulted in text-based compromises, which conveyed divisions to Khartoum and were a barrier to implementation. Through 2004, the Council clearly demonstrated to Khartoum that there was insufficient will to implement wider economic sanctions. Even once sanctions were implemented, the result was an arms embargo on the part of Sudan, and targeted sanctions on four individuals—only two of whom were from the government of Sudan—none of which were meaningfully enforced (Dörfler 2019). That decisions are shaped by social and relational mechanisms like legitimation practices does not necessarily mean that decisions are better in any normative sense or even implementable—the goal here is not to assess the implementation of decisions—rather to provide a fuller account of how decisions were made. The drive towards legitimation was instrumental in states taking the positions that they took and enables a clearer understanding of how decisions on sanctions were enabled.

## Bibliography

AFP. 2006. *Russia, China oppose sanctions on Sudanese offices over Darfur*: US, 17 April.

All Africa Global Media. 2006. *UN: Leaked names highlight Sudan's failure to pro-tect strong UN force; Sanctions needed to combat continuing attacks in Darfur*, 27 February.

AP. 2004a. 'US Issues 3rd UN Sudan Draft Resolution, Presses for Vote'. *Sudan Tribune*, 29 July.

AP. 2004b. 'US Threatens Sanctions Against Sudan'. *The Age*, 3 July.

Arieff, Irwin. 2006a. 'Top Sudanese Officials Targeted for UN Sanctions'. *Reuters*, 22 February.

Arieff, Irwin. 2006b. 'UN Experts Call for Sanctions on Darfur Leaders'. *Reuters*, 11 January.

Author Interview. 2011a. *Author Interview with a Think Tank in New York*, 4 February, New York.

Author Interview. 2011b. *Author Interview with P5 Diplomat*, 12 September, Paris.

BBC. 2004. 'US Demands Urgent Darfur Action'. *BBC*, 29 July.

Bellamy, Alex J. 2005. 'Responsibility to Protect or Trojan Horse? The Crisis in Darfur and Humanitarian Intervention after Iraq'. *Ethics and International Affairs* 19(2): 31–53.

Boulden, Jane and Andrea Charron. 2009-10. 'Evaluating UN Sanctions: New Ground, New Dilemmas, and Unintended Consequences'. *International Journal* 65(1): 1–11.

Brunk, Darren. 2008. 'Dissecting Darfur: Anatomy of a Genocide Debate'. *International Relations* 22(1): 25–44.

Charron, Andrea and Clara Portella. 2015. 'The UN, Regional Sanctions and Africa'. *International Affairs* 91(6): 1369–1385.

Clough, Michael. 2005. *Darfur: Whose Responsibility to Protect?* Human Rights Watch. https://www.hrw.org/legacy/wr2k5/darfur/darfur.pdf

Committee on Foreign Relations. 2004. *The Current Situation in Sudan and the Prospects for Peace*. United States Senate, 108th Congress, 9 September. Available at http://www.gpo.gov/fdsys/pkg/CHRG-108shrg97822/html/CHRG-108shrg97822.htm.

Daily Star. 2004. 'UN Adopts Softened Sudan Resolution; US-Backed Proposal Threat-ening Khartoum with Economic Sanctions', 31 July.

Danforth, Ambassador John. 2004a. *Remarks by Ambassador John Danforth on the Situation in the Sudan*. New York: U.S. Department of State Archive. Available at http://2001-2009.state.gov/p/af/rls/rm/34839.htm.

Danforth, Ambassador John. 2004b. *Remarks by Ambassador John Danforth, US Rep-resentative of the United Nations, on the Resolution Addressing the Situation in the Sudan, at the Security Council Stakeout, July 29 2004*. New York: U.S. Mission to the United Nations. Available at http://www.archive.usun.state.gov/press_releases/20040729_141.html.

de Waal, Alex. 2007. 'Darfur and the Failure of the Responsibility to Protect'. *International Affairs* 83(6): 1039–1054.

Dinmore, Guy, Andrew England, and Mark Turner. 2006. 'US Opposes British List for Sudan War Crime Sanctions'. *Financial Times*, 6 April.

Dörfler, Thomas. 2019. *Security Council Sanctions Governance: The Power and Limits of Rules.* Routledge Research on the United Nations. Abingdon and New York: Routledge.

Eckert, Sue. 2016. 'The Role of Sanctions'. In *The UN Security Council in the 21st Century*, eds. Sebastian Von Einsiedel, David M. Malone, and Bruno Stagno Ugarte. Boulder and London: Lynne Rienner Publishers.

El-Tablawy, Takek. 2004. 'US Waters Down Sudan Resolution, but Opposition Lingers'. *The Associated Press.*

Elliott, Kimberly Ann. 2009-10. 'Assessing UN Sanctions After the Cold War'. *International Journal* 65(1): 85–97.

Elster, Jon. 1995. 'Strategic Uses of Argument'. In *Barriers to Conflict Resolution*, eds. Kenneth J. Arrow, Robert H. Mnookin, Lee Ross, Amos Tversky, and Robert B. Wilson. New York and London: W.W. Norton & Company.

Elster, Jon. 1998. 'Deliberation and Constitution Making'. In *Deliberative Democracy*, ed. Jon Elster. Cambridge: Cambridge University Press.

Farrall, Jeremy. 2009. 'Should the United Nations Security Council Leave It to the Experts? The Governance and Accountability of Un Sanctions Monitoring'. In *Sanctions, Accountability and Governance in a Globalised World*, eds. Jeremy Farrall and Kim Rubenstein. Cambridge: Cambridge University Press.

Gamel, Kim. 2004. 'New US Draft Security Council Resolution on Sudan Still Faces Opposition'. *Sudan Tribune*, 28 July.

Gehring, Thomas and Thomas Dörfler. 2019. 'Constitutive Mechanisms of UN Security Council Practices: Precedent Pressure, Ratchet Effect, and Council Action Regarding Intrastate Conflicts'. *Review of International Studies* 45(1): 120–140.

Gehring, Thomas, Christian Dorsch, and Thomas Dörfler. 2019. 'Precedent and Doctrine in Organisational Decision-Making: The Power of Informal Institutional Rules in the United Nations Security Council's Activities on Terrorism'. *Journal of International Relations and Development* 22(1): 107–135.

Gifkins, Jess. 2016. 'R2P in the UN Security Council: Darfur, Libya and Beyond'. *Cooperation and Conflict* 51(2): 148–165.

Glanville, Luke. 2009. 'Is "Genocide" Still a Powerful Word?'. *Journal of Genocide Research* 11(4): 467–486.

Global Policy Forum. 2005. 'Draft Security Council Resolution on Sudan'. Accessed 18 May 2017. Available at https://www.globalpolicy.org/component/content/article/206/39640.html.

Grünfeld, Fred and Wessel Vermeulen. 2009. 'Failures to Prevent Genocide in Rwanda (1994), Srebrenica (1995), and Darfur (since 2003)'. *Genocide Studies and Prevention* 4(2): 221–237.

Grünfeld, Fred, Wessel Vermeulen, and Jasper Krommendijk. 2014. *Failure to Prevent Gross Human Rights Violations in Darfur: Warnings to and Responses by International Decision Makers (2003-2005).* Leiden: Brill Nijhoff Publishers.

Guéhenno, Jean-Marie. 2015. *The Fog of Peace: A Memoir of International Peacekeeping in the 21st Century.* Washington D.C.: The Brookings Institution.

Hamilton, Rebecca. 2011a. *Fighting for Darfur: Public Action and the Struggle to Stop Genocide.* New York: Palgrave Macmillan.

Hamilton, Rebecca. 2011b. 'Inside Colin Powell's Decision to Declare Genocide in Darfur'. *The Atlantic*, 17 August.

Heinze, Eric A. 2007. 'The Rhetoric of Genocide in U.S. Foreign Policy: Rwanda and Darfur Compared'. *Political Science Quarterly* 122(3): 359–383.

Hoge, Warren. 2005. 'France Asking UN to Refer Darfur to International Court'. *The New York Times*, 24 March.

Holslag, Jonathan. 2008. 'China's Diplomatic Manoeuvring on the Question of Darfur'. *Journal of Contemporary China* 17(54): 71–84.

Hurd, Ian. 2002. 'Legitimacy, Power, and the Symbolic Life of the UN Security Council'. *Global Governance* 8(1): 35–51.

Hurd, Ian. 2007. *After Anarchy: Legitimacy and Power in the United Nations Security Council*. Princeton: Princeton University Press.

Interfax. 2005. *Russia against Arms Embargo against Sudanese Govt*. Moscow: Interfax Information Services.

Johnstone, Ian. 2011. The Power of Deliberation: International Law, Politics, and Organizations. Oxford: Oxford University Press.

Keane, Fergal. 2005. *John Danforth Interview Transcript*. BBC. Available at http://news.bbc.co.uk/2/hi/programmes/panorama/4647211.stm.

Kirkup, James. 2004. *Up to 50,000 refugees dead in Sudan crisis. The UN's response? Delay, cowardice and inaction*, 31 July.

Lederer, Edith M. 2004. 'US Softens Threat of Possible Oil Sanctions against Sudan to Try to Get UN Approval for Tough Resolution to End Darfur Conflict'. The Associated Press, 14 September.

Lederer, Edith M. 2005. 'Deadlocked Council under Mounting Pressure to Adopt New Sudan Resolution Authorizing Force, Sanctions and Prosecutions'. The Associated Press, 19 March.

Lederer, Edith M. 2006. 'In Blow to Britain and US, Russia and China Will Oppose Sanctions against Sudanese Blocking Peace in Darfur'. Associated Press, 17 April.

Leopold, Evelyn. 2004. 'US Calls for Friday Vote on Darfur UN Measure'. *Sudan Tribune*, 29 July.

Leopold, Evelyn. 2005a. 'Annan Calls Emergency Sudan Session of UN Council'. Reuters, 7 March.

Leopold, Evelyn. 2005b. 'UN Council Delays Sudan Sanctions by Three Months'. Reuters, 6 July.

Leopold, Evelyn. 2006a. 'US against Sanctions on Sudan Officials—Diplomats'. Reuters, 5 April.

Leopold, Evelyn. 2006b. 'US Readies Darfur Sanctions Vote at the UN This Week'. Reuters, 24 April.

Lynch, Colum. 2004. 'US Seeks UN Pressure on Sudan'. *The Washington Post*, 23 July.

Mohammed, Arshad. 2005. 'US Renews Push for Oil Sanctions on Sudan'. Reuters, 1 February.

Reuters. 2004. 'UN Council Divided over Sanctions on Sudan'. *Daily Times*, 11 July.

Reuters. 2005. 'Highlights of New US Draft on Sudan at UN'. Reuters, 15 February.

Security Council Report. 2008. *Security Council Action Under Chapter VII: Myths and Realities*. New York. Available at http://www.securitycouncilreport.org/atf/cf/%7B65BFCF9B-6D27-4E9C-8CD3-CF6E4FF96FF9%7D/Research%20Report%20Chapter%20VII%2023%20June%202008.pdf.

Seymour, Lee J.M. 2014. 'Let's Bullshit! Arguing, Bargaining and Dissembling over Darfur'. *European Journal of International Relations* 20(3): 1–25.

Stedjan, Scott and Colin Thomas-Jensen. 2010. 'The United States'. In *The International Politics of Mass Atrocities: The Case of Darfur*, eds. Paul D. Williams and David R. Black. London and New York: Routledge.

Taylor, Ian. 2010. 'The People's Republic of China'. In *The International Politics of Mass Atrocities: The Case of Darfur*, eds. Paul D. Williams and David R. Black. London and New York: Routledge.

Tierney, Dominic. 2005. 'Irrelevant or Malevolent? UN Arms Embargoes in Civil Wars'. *Review of International Studies* 31(4): 645–664.

Totten, Samuel 2010. 'Saving Lives in Darfur, 2003–06?: Lots of Talk, Little to No Action'. In *The World and Darfur: International Response to Crimes against Humanity in Western Sudan*, 2nd ed., ed. Amanda F. Grzyb. Montreal and Kingston: McGill-Queen's University Press.

Traub, James. 2010. *Unwilling and Unable: The Failed Response to the Atrocities in Darfur*. New York. Available at http://www.responsibilitytoprotect.org/GCR2P_UnwillingandUnableTheFailedResponsetotheAtrocitiesinDarfur.pdf.

Turner, Mark. 2005. 'US Urges Creation of 10,000-Strong Sudan Peace Force'. *Financial Times*. 15 February.

Tzanakopoulos, Antonios. 2014. 'Strengthening Security Council Accountability for Sanctions: The Role of International Responsibility'. *Journal of Conflict & Security Law* 19(3): 409–426.

United Nations. 1945. *Charter of the United Nations*. San Francisco: United Nations.

United Nations. 2004a. *5015th Meeting United Nations Security Council*. New York, 30 July, S/PV.5015.

United Nations. 2004b. *5040th Meeting of the United Nations Security Council*. New York, 18 September, S/PV.5040.

United Nations. 2004c. *Joint Communique between the Government of Sudan and the United Nations on the Occasion of the Visit of the Secretary General to Sudan 29 June - 3 July 2004*. Available at http://www.un.org/news/dh/sudan/sudan_communique.pdf.

United Nations. 2004d. *Report of the Secretary-General pursuant to paragraphs 6 and 13 of Security Council resolution 1556 (2004)* New York. Available at https://documents-dds-ny.un.org/doc/UNDOC/GEN/N04/474/12/PDF/N0447412.pdf?OpenElement.

United Nations. 2004e. *Resolution 1556*. 30 July, S/RES/1556.

United Nations. 2004f. *Resolution 1564*. 18 September, S/RES/1564.

United Nations. 2005a. *5153rd Meeting United Nations Security Council*. New York, 29 March, S/PV.5153.

United Nations. 2005b. *Report of the International Commission of Inquiry on Darfur to the United Nations Secretary-General*. Geneva, 25 January. Available at http://www.un.org/news/dh/sudan/com_inq_darfur.pdf.

United Nations. 2005c. *Resolution 1591*. New York, 29 March, S/RES/1591.

United Nations. 2006. *5423rd Security Council Meeting*. New York, 25 April, S/PV.5423.

United Nations. 2022. *UN Security Council Meetings & Outcomes Tables* Dag Hammarskjöld Library. Available at https://research.un.org/en/docs/sc/quick/meetings/2022.

US Embassy Khartoum. 2006. *Sudan/China: UNSC vote on Darfur shows limits of friendship*. Wikileaks. Available at https://wikileaks.org/plusd/cables/06KHARTOUM996_a.html.

Vine, Alex. 2007. 'Can UN Arms Embargoes in Africa Be Effective?'. *International Affairs* 83(6): 1107–1122.

Voeten, Erik. 2005. 'The Political Origins of the UN Security Council's Ability to Legitimize the Use of Force'. *International Organization* 59(3): 527–557.

Wenqi, Zhu and Leng Xinyu. 2016. 'China in the Security Council'. In *The UN Security Council in the Twenty-First Century*, eds. Sebastian von Einsiedel, David M. Malone, and Bruno Stagno Ugarte. Boulder: Lynne Rienner.

Williams, Paul D. 2006. 'Military Responses to Mass Killing: The African Union Mission in Sudan'. *International Peacekeeping* 13(2): 168–183.

Williams, Paul D. 2010. 'The United Kingdom'. In *The International Politics of Mass Atrocities: The Case of Darfur*, eds. Paul D. Williams and David R. Black. London and New York: Routledge.

Williams, Paul D. and Alex J. Bellamy. 2005. 'The Responsibility to Protect and the Crisis in Darfur'. *Security Dialogue* 36(1): 27–47.

Wood, Michael C. 1998. 'The Interpretation of Security Council Resolutions'. In *Max Planck Yearbook of United Nations Law*, eds. Jochen A. Frowein and Rudiger Wolfrum. London: Kluwer Law International.

Wuthnow, Joel. 2010. 'China and the Processes of Cooperation in UN Security Council Deliberations'. *Chinese Journal of International Politics* 3(1): 55–77.

Xinhua. 2004. 'UK Says Most Security Council Members Oppose Tough Sanctions on Sudan'. London, 20 August.

Xinhua. 2005. 'China Opposes Economic Sanctions against Sudan's Darfur Crisis'. 17 March.

# 6

# The UN Security Council's First Referral to the ICC

*The Council is not a simple aggregate of fifteen separate interests ... The Council also has a semblance of solidarity ... and contains norms that keep in line even the most independent-minded.*

(Barnett 2002: 11)

The Security Council's referral of the situation in Darfur to the International Criminal Court (ICC) in 2005 was uniquely contentious among the Council's negotiations on Darfur. There were key national interests at stake for the US government, which supported international criminal justice on the proviso that it could prevent any investigation of US personnel. This meant that the US was wary of any potential ICC jurisdiction for US citizens. During the drafting of the Rome Statue, which founded the ICC, the US failed to convince other states that the ICC should be subservient to the UN Security Council, which would have enabled the US to veto decisions. The compromise was that the Security Council can defer an investigation or prosecution under Article 16 of the Rome Statute for one year at a time, and that the Security Council can refer cases to the ICC under Article 13(b) of the Rome Statute, which established a formal, but not subservient, relationship between the two bodies. Despite not being a party to the Rome Statute, it was plausible that US citizens could be investigated for crimes committed on the territory of states which were party to the Rome Statute. US opposition to the ICC therefore predates the Darfur referral, and the negotiations towards the Darfur referral hinged on gaining US acquiescence. The referral substantiated the relationship between the Security Council and the ICC—via its first referral of a case—so it was contentious as it established a precedent. The national interests in play for the US in negotiating the referral of Darfur to the ICC were stronger than the various national interests for permanent members in opposing agenda-setting, sanctions, or peacekeeping in Darfur.

The puzzle is why did the US—with a long-standing opposition to the ICC—allow this referral to happen? The US had the capacity to individually block this decision via a veto. Its statement after the vote explained that 'we have not dropped, and indeed continue to maintain, our long-standing and firm objections and

*Inside the UN Security Council*. Jess Gifkins, Oxford University Press. © Jess Gifkins (2023).
DOI: 10.1093/oso/9780192869029.003.0007

concerns regarding the ICC' (United Nations 2005a: 3). Yet the US enabled this precedent-setting resolution to pass. One possible answer is that the US achieved what it wanted in the resolution in the form of an exemption from the jurisdiction of the ICC for states which are not party to the Rome Statute (excluding Sudan). If this is the primary answer then perhaps the Security Council did little to shape the actions of the US, in line with Mearsheimer's contention that institutions only matter 'on the margins' (1994/1995: 7). However, it was clear from the beginning of the negotiations on accountability in Darfur that the US was not even willing to consider an ICC referral. The US opposed the idea of legitimizing the ICC via an endorsement from the UN Security Council, it sought support for an alternative ad hoc tribunal, and it came very close to vetoing the ICC referral. The US only decided to abstain on the day of the vote. These initiatives show a profound unwillingness to legitimize the ICC by referring the case of Darfur to it. At the eleventh hour, 'Washington found that the cost of a veto would be higher than the cost of abstaining' (Johansen 2006: 330). This 'cost' was to the legitimacy and reputation of the US and its relationships within the UN Security Council, particularly with European states who were at the forefront of the push for an ICC referral. Civil society groups on Darfur in the US were strong by this stage, along with advocates in Congress, which increased the costs of vetoing an ICC referral for the US, however, the Save Darfur movement took the vaguer line of an 'end to impunity' in Darfur, rather than specifically championing the ICC as the goal (Hamilton 2011: 68). A key factor which enabled the resolution to pass, and prevented a US veto, were external legitimation practices and the desire for the US to increase rather than decrease its social status of legitimacy.

There are a variety of existing rationalist and constructivist explanations for the US enabling the referral of Darfur to the ICC, in addition to the argument presented in this chapter on legitimation practices. The US achieved an exemption for individuals from states that are not party to the Rome Statute which is a strategic concession, and this is the most common explanation given for why the US decided to abstain (Condorelli and Ciampi 2005; Cryer 2006; Heyder 2006). There are also potential rationalist arguments around the reputational costs for the US if it had vetoed the resolution. However, as others have identified, there are also constructivist explanations for the US' behaviour in relation to the ICC in general (Fehl 2009), and specifically related to the referral of the situation in Darfur (Mills and Lee 2007). Of all the situations considered in this book, the ICC referral was the one which invoked the strongest national interests for a Security Council member. Yet even here there is evidence that the US was constrained in achieving its own interests of delegitimizing the ICC due to external legitimation practices, even though it was within the power of the US to individually block this decision.

The Security Council's first referral of a case to the ICC was significant for three key reasons, all connected to the Security Council legitimating the ICC. First, the precedent value of a first referral. This step substantiated the relationship between the Security Council and the ICC, which had only existed on paper up until this point. Precedent is also important for the practices of the Security Council, as it is easier to find consensus on language that has been used in past decisions, so the first referral established a set of 'agreed language' for use in future referrals (Gifkins 2016; Werner 2017). The caveats used in the Darfur referral were subsequently replicated in the Security Council's second referral to the ICC, which was the situation in Libya (Dunne and Gifkins 2011: 524). Second, three of the five permanent members—the US, China, and Russia—were not parties to the Rome Statute, which can be seen to position those three states as above the law (Arbour 2014: 198–199). Each of these states had the capacity to veto the referral, so the fact that they decided not to amounted to a tacit endorsement of the ICC as an institution. Third, the US had engaged in considerable obstruction of the ICC by the time of the Darfur referral, so US acquiescence of the referral was particularly significant. Each of these three reasons meant that there was considerable contestation over the legitimacy of the ICC, and its relationship with the Security Council, involved in the referral of the situation in Darfur.

As such, the Security Council referral of the situation in Darfur to the ICC hinged on competing visions of legitimacy. Since it was the first situation that the Security Council had referred to the ICC, it represented a new stage in the relationship between the ICC and the Council and the legitimation of the ICC. Throughout the negotiations, the Bush administration was keen to avoid any legitimation of the ICC. Interviewees involved in the negotiations were clear that the politics of these negotiations were primarily about the ICC rather than about Darfur (Author Interview 2011a, 2011b). In contrast to this priority, however, the US had also established itself as a prominent advocate on Darfur and had determined that the situation in Darfur constituted genocide.[1] While a US veto was a real possibility on the draft referral, the US was constrained because the conflict was Darfur rather than a situation it was less invested in. As many news article argued at the time, there would have been substantial legitimacy costs for the US to veto a referral of Darfur to the ICC (see for example Bravin and Paltrow 2005; Hoge 2005; Wadhams 2005). As such, while the negotiations were primarily focused on the politics of the ICC, the former Special Advisor to the UN Secretary General on the responsibility to protect Edward Luck was right when he argued that the US' eventual decision to abstain on the referral was all about Darfur (Author Interview 2011d). Despite not wanting to legitimize the ICC, there would have been legitimacy costs to the US to veto the draft which would have given the appearance of supporting impunity in Sudan.

This chapter proceeds in four parts. Since the key obstacle to the referral was the US the first section is an overview of the US' relationship with the ICC. As others

---

[1] See Chapter 5 for further discussion on this point.

have highlighted, China and Russia were not as resistant to the ICC referral as the US, so the key challenge in these negotiations was overcoming US resistance (Grünfeld, Vermeulen and Krommendijk 2014: 198). The second section charts the Security Council's early engagement with the ICC, before the Darfur referral. Although the referral of Darfur occurred in 2005, the Security Council began to pass resolutions about the ICC in 2002, and the referral of Darfur only makes sense within the context of these early negotiations which were structured around US opposition to the ICC. The third section explains the steps towards resolution 1593 which referred the situation in Darfur to the ICC. The final section analyses the US' decision to abstain and enable this resolution. Drawing on interview data, news archive material, and secondary sources, this chapter reconstructs the Security Council's negotiations towards the Darfur referral. This chapter shows that the US was constrained in furthering its interest of delegitimizing the ICC. The referral of Darfur solidified the previously hypothetical relationship between the ICC and the Security Council. External legitimation practices of the US as an actor—both domestically and from other states—led the US to abstain rather than veto this resolution.

## The United States' Relationship to the International Criminal Court

A key priority for the US was to retain control over the then nascent International Criminal Court by requiring Security Council approval for individual cases, thus ensuring that the US had veto over which cases are investigated. The US had supported earlier ad hoc tribunals for the former Yugoslavia and for Rwanda but was wary of the idea of an independent system of international criminal justice that was divorced from the Security Council. This was such a high priority for the US that prominent scholar William Schabas has argued that US opposition to the ICC is 'all about the Security Council' (2004). Security Council control is clearly a high priority for the US, but this is underpinned by what Security Council control can enable: immunity for the activity of US citizens abroad. The US is willing to accept the ICC investigating nationals from non-party rouge states without a referral from the Security Council yet is fundamentally opposed to its own nationals coming under the jurisdiction of the ICC (Scheffer 2001: 65). Former US Ambassador David Scheffer, who led the US delegation during the negotiations towards the Rome Statue, stated this bluntly: 'The central issue confronting the United States government with respect to the ICC is the risk that the Court may seek to investigate, obtain custody of, and ultimately prosecute a US service member or US Government official in connection with that individual's official duty' (2001: 87). This issue is of such a high priority for the US for two reasons. First, the US occupies the role of hegemon with a greater involvement in military activity abroad than other states (Mills and Lee 2007: 506–507). Second, the US has a particular cultural affiliation with the primacy of sovereign consent (Ralph 2007). In this sense, the primary aim is immunity for US personnel, and control of the ICC via the UN Security Council is a means of achieving this aim.

The US participated in the negotiations towards the Rome Statute with the hope of establishing a system of international criminal justice that was subservient to the UN Security Council. Under President Clinton, the US signed the Rome Statute, to enable its continued influence in the Preparatory Commission meetings in 2000 towards the establishment of the ICC. During these negotiations, the US tried to secure exemptions for troops who serve in operations that are authorized by the UN Security Council (Ralph 2007: 163). On the question of Security Council control of the ICC, the US lost this debate during the drafting of the Rome Statute (Schabas 2004). Failing to secure this concession, the subsequent Bush administration increased the US' hostility to the ICC and announced that it would not ratify the Rome Statute.

A key dimension of the US relationship with the Rome Statute is the controversial interpretation the US holds of Article 16. This provision enables to the Security Council to defer an ICC investigation or prosecution for a renewable period of twelve months, if it adopts a resolution under Chapter VII of the UN Charter (Rome Statute of the International Criminal Court 1998). During the drafting of the Rome Statute, this provision was intended to apply to specific situations (i.e. conflicts), rather than provide broad immunity for types of actors (Jain 2005). Yet, in line with its key priorities, the US interpretation of Article 16 is that it can be used to grant immunity for a set of actors. As such, much of the negotiation within the Security Council in relation to the ICC has focused on Article 16, as the US has attempted to use this provision to facilitate immunity for US personnel.

## Early Days of the ICC and US' Attempts to Secure Exemptions

The Bush administration wanted exemptions for US personnel in place before the Rome Statute came into effect in July 2002. By the end of June, the US had not been able to secure this, so it forced the issue by vetoing the renewal of a peacekeeping operation in Bosnia and Herzegovina. The US was clear in its statement that the veto related to the failure to reach a decision on immunity for non-signatories to the Rome Statute, rather than to the peacekeeping operation at hand (United Nations 2002a: 2–3). By doing so, the US made it clear that it was willing to veto resolutions if it did not obtain the immunity it wanted. The Security Council held an open meeting following the veto (meaning non-Security Council members could attend) and more than three dozen statements were made, the vast majority of which were critical of the US' proposal to use Article 16 to create a blanket deferral for twelve months for troops in peacekeeping operations (United

Nations 2002b). However, vetoing a resolution on peacekeeping, and threatening to veto further peacekeeping decisions, gave the US considerable leverage in these negotiations.

Two days later, the Security Council passed resolution 1422, which provided this immunity. The resolution passed unanimously and without statements, but it is clear that this was a response to the US veto. With the US blocking peacekeeping and threatening to veto more peacekeeping resolutions, Security Council members agreed to a resolution which provided immunity to officials working in UN established or authorized operations, from states which are not signatories to the Rome Statue (United Nations 2002c). Drawing from Article 16 of the Rome Statute, this decision offered immunity for a twelve-month period. An early draft of this resolution allowed for automatic renewal, unless the Security Council decided not to renew it (Ralph 2007: 165). This would have meant that the veto would only apply to the termination of the decision, not to its continuation, which would have given the US indefinite access to this immunity. This provision proved a bridge too far for other states, however, and was removed from the final version, which meant that the US required annual renewal of this decision to maintain immunity. Having achieved the desired immunity for US officials for twelve months, the US allowed the Bosnia and Herzegovina peacekeeping operation to be renewed the same day. In this process, the US made it clear that immunity for its citizens was such a high priority that it was willing to veto unrelated resolutions where necessary to achieve this goal.

In 2003, the US again lobbied for a renewal of the twelve-month exemption for individuals from non-signatories to the Rome Statute. It achieved this in resolution 1487, but resistance to these annual exemptions had increased. The UN Secretary-General argued that when the ICC was in its early months of operation the previous year, he could understand the rationale for this decision despite his view that Article 16 was only meant to apply to situations. However, he was clear that the scenario the resolution guarded against—where a peacekeeper of a non-signatory state was accused of crimes which fitted the purview of the ICC and its home country was not willing to investigate—was highly unlikely, and that this should not become an annual renewal (United Nations 2003a: 2–3). While the first resolution of this kind in 2002 was unanimous, in 2003 there were abstentions from France, Germany, and Syria, also indicating increased resistance, especially given that the vast majority of Security Council resolutions are unanimous. At an open meeting, states spoke about their dissatisfaction with this resolution and many said that it was inconsistent with the intentions of Article 16 of the Rome Statute (United Nations 2003a).

Not satisfied with this blanket immunity for peacekeepers, or perhaps to create additional precedent in case further renewals could not be secured, the US added a clause to a peacekeeping resolution on Liberia to grant immunity from prosecution to personnel from non-signatory states who were part of that peacekeeping operation (United Nations 2003c). Resolution 1497 carved out an exemption whereby peacekeepers from states which were not signatories of the Rome Statute would not come under the jurisdiction of the ICC. This resolution was met with abstentions from France, Germany, and Mexico, all three of whom expressed concerns about the compatibility of this decision with international law, and argued for a paragraph-by-paragraph vote on the resolution, so that the question of immunity could be isolated from the wider resolution on peacekeeping in Liberia (United Nations 2003b). Arguably, this resolution at least had a closer alignment with the Rome Statute, as the deferral applied to a specific situation that had been deemed to be a threat to international peace and security, while earlier blanket resolutions 1422 and 1487 did not meet that criteria (Jain 2005). The US demonstrated that it wanted to establish precedents for immunity via any means possible within the Security Council.

The US' immunity provision was due to expire (if not renewed) in June 2004; however, by April 2004, there was widespread media attention on alleged torture and dehumanization at Abu Ghraib prison in Iraq. The potential consequences of US immunity were no longer hypothetical for other Security Council members. Early in the negotiations towards renewal, four states said they would abstain and another two were considering abstaining, which left the minimum number of affirmative votes if all other states voted in favour, which was not a given (Lederer 2004b). In addition to the possibility of not reaching the required minimum, China announced that it was considering vetoing the US' draft. Interestingly, although China said that this was a reaction to the Abu Ghraib allegations, diplomats in the Security Council believed it to be a retaliation against US support for Taiwan to gain observer status at the World Health Assembly (Leopold 2004b; Lynch 2004). The US recognized it was losing traction on the idea of annual renewals and suggested a compromise whereby 2004 would be the last annual renewal (Leopold 2004a). In the context of the Abu Ghraib allegations, even this compromise was not enough. In the end, eight Security Council members said that they would abstain on the resolution, including China, which left a possible affirmative vote of seven, which would have been below the legal threshold of nine (Lederer 2004a). The US realized that the draft would not pass and abandoned it, meaning that the blanket immunity lapsed in 2004.

These attempts by the US to use the Security Council to negotiate immunities for US personnel set the context for the negotiations towards the Darfur referral. By the time the Security Council debated how best to end impunity in Darfur, there had been three years of the US vehemently negotiating exemptions for non-party states into blanket resolutions on peacekeeping and into resolutions on specific peacekeeping operations. The US had also demonstrated that this was such a high priority that it was willing to veto drafts to secure this objective. Another indicator of how high a priority this was for the US were the bilateral non-surrender agreements that it negotiated with states, one at a time. The US attempted to establish bilateral agreements with states across the world; however, even when threatened with the withdrawal of foreign aid, many states refused because they valued the ICC as an institution or because they valued the rule of law (Kelley 2007). Interestingly, this move went beyond the potential practical consequences of foreseeable scenarios, and was another attempt by the US to delegitimize the ICC (Kelley 2007: 576). Following on from extensive attempts to secure immunity for US personnel, the US had two key priorities when approaching the Security Council negotiations on Darfur and the ICC: delegitimizing the ICC—ideally by not referring the situation in Darfur to the ICC—and establishing further precedents on immunity for US personnel.

## The UN Security Council, the ICC, and Darfur

The politics of the ICC became inextricably linked to Darfur following the report of the International Commission of Inquiry on Darfur (ICID) in January 2005, which called on the Security Council to refer the situation in Darfur to the ICC. Since Sudan was not a party to the Rome Statute, a referral by the Security Council was an alternative means to bring the crimes in Darfur under the ICC's jurisdiction. As discussed in Chapter 5, the International Commission of Inquiry on Darfur was established by the Security Council under resolution 1564 in 2004. The mandate for the ICID was to investigate violations of international humanitarian law and human rights law, to determine whether acts of genocide had occurred, and to identify perpetrators with the intention of holding perpetrators to account (United Nations 2004: 3–4). While not specifically tasked with identifying the means of accountability, it is clear that questions of accountability were within the remit of the ICID.[2] The Commission found that the situation in Darfur met the criteria for referral to the ICC with widespread and systematic alleged crimes, including war crimes and crimes against humanity, which the Sudanese government was unwilling to investigate. The ICID report was also clear that the ICC was the most appropriate court to investigate these crimes, and, foreshadowing counterarguments by the US, it argued that it would be both cheaper and quicker to use the ICC than to establish an ad hoc tribunal (United Nations 2005b). Indeed, the ICID concluded that a Security Council referral to the ICC was 'the only credible way of bringing alleged perpetrators to justice' (United Nations 2005b: 146). Since the ICID was mandated by the Security Council, and since it made such a compelling and forceful argument that the Council should refer the situation in Darfur to the ICC, the ICC became central to the Council's negotiations over how to end impunity in Darfur.

---

[2] It has also been suggested that the ICID was a specific strategy by states that supported ICC investigation of Darfur to generate momentum for a referral to the ICC (Grünfeld, Vermeulen and Krommendijk 2014: 189).

From the outset, the question of referring Darfur to the ICC posed real challenges for the Bush administration. By 2005, it had established itself as one of the most vocal and active states criticizing violence in Darfur and had described the violence as genocide. This activism made it politically difficult to veto a referral. Yet it did not want to legitimize the ICC by referring a case to it, which created a bind for the US. Before the ICID released its report, the US recognized the need for accountability in Darfur, but a US official stated, 'we are studying the options and the ICC is not one of them' (Bravin and Paltrow 2005). Once the ICID report was released, the ICC became a key possibility for accountability, and the avenue favoured by other states. Days after the release of the ICID report, Pierre-Richard Prosper, the US' ambassador-at-large for war crimes stated bluntly, 'Our position is well-known ... we don't want to be party to legitimizing the ICC' (Leopold 2005c). US administration officials recognized that not vetoing a referral would 'effectively give the ICC a US stamp of approval and fatally undercut efforts to marginalise it' (Bosco 2014: 111). This put the US in a difficult position from the outset. The Security Council–mandated ICID made a clear and compelling argument that the ICC was the most appropriate forum for investigating crimes committed in Darfur, but the Bush administration had a fundamental opposition to legitimating the ICC by referring a case to it.

In mid-February 2005, the US circulated a draft resolution which engaged with some of the other political challenges Sudan faced at the time. The draft would have authorized a UN peacekeeping operation in Sudan to monitor the Comprehensive Peace Agreement (CPA) between the north of Sudan and the south of Sudan, and established a Sanctions Committee to monitor the arms embargo on Darfur (Global Policy Forum 2005).[3] It noted the findings of the ICID, but dodged the issue of accountability by declaring that perpetrators 'must be brought to justice through internationally accepted means' without any indication of how this would be done (Global Policy Forum 2005). By the start of 2005, ninety-seven states had ratified the Rome Statute, and of the fifteen states on the UN Security Council at the time, nine were parties to the Rome Statute.[4] Deferring the question of accountability was not acceptable to other members of the Security Council. Also, in mid-February, Robert Zoellick was appointed as the US' Deputy Secretary of State, assisting Condoleezza Rice. His recommendation on Sudan was to split the draft into three separate resolutions, so the Council could proceed on the points where there was agreement (Hamilton 2011: 62–63).[5] This recommendation was followed, and the Security Council passed resolution 1590,

---

[3] This draft is also discussed in Chapter 5 in relation to sanctions.

[4] These were Argentina, Benin, Brazil, Denmark, France, Greece, Romania, the United Kingdom, and Tanzania. The Philippines signed the Rome Statute in 2000 but did not ratify until 2011.

[5] The Security Council regularly approached the concurrent conflicts in Darfur and the southern region of Sudan (now South Sudan) with a compartmentalized and sequential approach (for discussion see Stedjan and Thomas-Jensen 2010).

authorizing peacekeeping to monitor the CPA, and resolution 1951, which established a Sanctions Committee and Panel of Experts to monitor the arms embargo and to designate individuals for targeted sanctions. The question of impunity was by far the most contentious of the three issues and was tackled separately.

There was a flurry of bilateral diplomacy throughout January and February as France sought support for a referral of Darfur to the ICC and the US sought support for an ad hoc court. France engaged with other Council members and there was broad support for an ICC referral. The US acknowledged that an international court was necessary to challenge impunity, rather than domestic courts in Sudan, but it sought any alternative that would avoid legitimizing the ICC. An internal memo from the US War Crimes Office from this period outlines the problem very clearly, as seen from a US perspective: 'We do not want to be confronted with a decision on whether to veto an ICC resolution [on Darfur] in the Security Council' (Hamilton 2011: 58). To circumvent momentum towards an ICC referral, the US instead proposed an ad hoc tribunal to be led by the UN and the African Union based in Arusha, Tanzania, using facilities that were being used for the International Criminal Tribunal for Rwanda. The US envisaged this as a permanent regional alternative to the ICC (US Embassy Paris 2005). There was little support for this plan. In fact, it gained so little traction that after a meeting of Security Council members to discuss the ICC referral in mid-February, a diplomat reported, 'Nobody mentioned the American proposal' (Arieff 2005).

In an attempt to gain traction for the Court in Tanzania, the US put forward a series of arguments: that the US was willing to cover most of the costs (Leopold 2005a); that it would be cheaper than the ICC (a point disputed by the ICID report) (Hamilton 2011: 61); that it could start sooner than the ICC (also disputed by the ICID report); and that it could include crimes committed in 2001 and 2002 which predate the mandate of the ICC (Kelley 2005). Tanzania was clear that it would only support this mechanism if there was strong support within the African Union and the UN Security Council (US Embassy Dar es Salaam 2005). Since this alternative proposal never gained traction with other Security Council members, it was infeasible for Tanzania. After this, Nigeria, which was Chairperson of the African Union at the time, put forward a proposal for an African panel to investigate crimes in Darfur, which also didn't gain traction (Leopold 2005b). Neither of these alternatives found wide appeal with Security Council members— the majority of which were parties to the Rome Statute—and strong supporters of the ICC.

By mid-February 2005, it was clear that the proposal to refer Darfur to the ICC had enough support to pass if there was no veto. Twelve members of the Security Council supported the proposed referral, with only the US, China, and Algeria in opposition (Arieff 2005). The US and China each had the capacity to individually block the decision. Unusually for Security Council negotiations, there was no drive towards unity for this decision; the goal was simply to avoid the legal obstacle of a US veto (Author Interview 2011c). Another month passed with further negotiations to avoid a veto before France forced the issue by circulating a draft resolution referring the situation in Darfur to the ICC. By forcing the issue, France left the US a choice between 'validating a tribunal it strongly opposes or casting a politically awkward veto' (Hoge 2005). The US asked France to delay the resolution while it drafted an alternative text. The alternative US draft, however, included broad exemptions for nationals of states which were not party to the Rome Statute—which, accidentally, would have not only included US nationals, but also Sudanese nationals—and a provision that the case be funded by state parties to the Rome Statute rather than by UN members (Hamilton 2011: 66). The first provision was clearly not feasible and was amended so that Sudanese nationals were not excluded, and then both provisions became part of the final resolution.

It was the UK which brokered a deal to make the resolution possible. France led the draft until almost the end of the negotiations, but it was unwilling to endorse the US amendment exempting peacekeepers from non-party states (Author Interview 2011a). The amendment used language from resolution 1497 on Liberia. 'Previously agreed language' carries particular diplomatic weight in negotiations, as once agreement has been found it is difficult for states to object (for discussion see Gifkins 2016; Werner 2017). Agreed language also functions as an internal legitimation practice—a standard negotiating technique for enhancing the legitimacy of a draft. France had abstained on the Liberian resolution due to concerns over the exemption, so France did not want to sponsor the Darfur referral once this language was included (Hamilton 2011: 67). France withdrew its sponsorship from the draft and the UK took over sponsorship to avoid political difficulties for France and to facilitate the US' amendment.

The US was able to achieve some of its priorities in these negotiations. The US extracted three key concessions, in exchange for abstaining on this resolution (for further discussion on the concessions see Condorelli and Ciampi 2005; Cryer 2006; Heyder 2006). By the time of the vote, the US had negotiated bilateral agreements with ninety-nine countries which exempted prosecution of US citizens by the ICC (United Nations 2005a: 4), but still wanted blanket immunity for peacekeepers and UN authorized peace enforcement operations. The key concession the US extracted was an exemption for individuals from non-party states (excluding Sudan). The second concession was to require funding to come from parties to the Rome Statute, rather than from the UN, which exempted the US from paying for ICC investigations into Darfur. This provision is difficult to reconcile with the Rome Statute, where Article 115 states an expectation that cases referred by the Security Council will be funded by the United Nations. As such, the lawfulness of resolution 1593 has been questioned (Cryer 2006; Elagab 2009; Schabas 2010). The third concession was an acknowledgement in the resolution of the existence of agreements under Article 98(2) of the Rome Statute, whereby bilateral agreements between states can circumvent the jurisdiction of the ICC over nationals of non-party states in situations where the ICC would otherwise have jurisdiction. These immunity agreements were a very high priority for the US—one that it was willing to prioritize above other foreign policy goals (Ralph 2007: 159). These concessions show that a high price was paid by advocates of the ICC to facilitate the Darfur referral.

The US did not, however, get all of what it wanted. As above quotes from US officials make clear, an ICC referral was not an outcome that the US was even willing to consider at the start of these negotiations, and the US fundamentally did not want to legitimize the ICC via a referral from the Security Council. This was such a high priority for the US that it offered to bear most of the costs of an ad hoc tribunal for Darfur—which was estimated to cost more annually than the entire costs of the ICC (Human Rights Watch 2005). There were other concessions that the US argued for which were not accepted by other Security Council members. In particular, the US wanted a provision that would allow non-party states to opt out of ICC jurisdiction for seven years (a provision that is available to state parties under Article 124 of the Rome Statute) (Leopold and Mohammed 2005). This blanket immunity was rejected by other Security Council members, and the eventual immunity granted to non-party states only applied to crimes committed within Sudan. The US also requested an acknowledgement in the resolution that not all Security Council members were party to the Rome Statute, and the US Ambassador reportedly called the UK Ambassador three times on the evening of the vote to request this (Hamilton 2011: 67). Despite not achieving all its goals, the concessions that the US did extract established a precedent. The operative paragraph granting immunity to non-party states was subsequently included in the Security Council's second referral to the ICC, which was the situation in Libya in 2011 (United Nations 2011). As such, the exemptions obtained by the US in the Darfur referral were precedent-setting, but by enabling the referral to occur, the US legitimated the ICC.

There is a caveat to the idea that the referral meant legitimizing the ICC while a veto would have meant delegitimizing the ICC. The concessions achieved by the US constrained the operation of the ICC, both financially and in terms of its jurisdiction, and set these concessions as precedents for future referrals (Arbour 2014). Speaking after the vote, many states expressed serious concerns with the immunity clause for non-party states, and for Brazil, a supporter of the ICC, this issue was so serious it decided to abstain from the vote, despite strongly supporting the referral of Darfur to the ICC (United Nations 2005a: 11). The Philippines framed the exemption of ICC jurisdiction as 'killing its credibility', while others such as France and Greece framed it as a necessary compromise (United Nations 2005a: 6 and 8–9). While there was some concern for states over the potential impacts of the exemption clause, and its precedent implications, it was a compromise that most Security Council members were willing to accept to secure the referral of Darfur to the ICC. As such, while the immunity clause was recognized by supporters of the ICC as not ideal, they also recognized that ICC investigation of crimes committed in Darfur was more important than the potential delegitimization inherent in the immunity provision and the provision which required only party states to the Rome Statute to cover the costs of the referral. The US did achieve some delegitimization of the ICC via its concessions, however, these limitations were acceptable for most Security Council members.

The US was in a difficult position, trying to reconcile its abstention (enabling the resolution to pass) with its opposition to the ICC. In its statement after the vote, Anne Patterson, the US Deputy Permanent Representative to the United Nations, stated bluntly that 'we do not agree to a Security Council referral of the situation in Darfur to the ICC'. She continued that 'we have not dropped, and indeed continue to maintain, our long-standing and firm objections and concerns regarding the ICC'. The US justified its abstention by stressing the need to end impunity and the exemption of nationals of non-party states (United Nations 2005a: 2–3). In the days that followed the vote, the US State Department issued a series of comments justifying its abstention and reiterating its opposition to the ICC. US Secretary of State Condoleezza Rice described Darfur as an 'extraordinary circumstance' and stressed the earlier US statements on genocide (US Department of State 2005a). Under Secretary of State for Political Affairs Nicholas Burns explained that 'This was a difficult decision for us, for obvious reasons, because of our longstanding position on the International Criminal Court. And our position on that Court has not changed as a result of this action' (US Department of State 2005b). After the referral of Darfur to the ICC, the US maintained its opposition to the Court, but it had also legitimated the Court by enabling a case to be referred by the Security Council.

## Interpreting the Referral of Darfur to the ICC

The referral of Darfur to the ICC was a very significant step for the ICC and its relationship to the Security Council. More so than the other negotiations considered in this book, there were strong national interests at stake for a permanent member. While China's interests in Sudan had some influence on decisions on sanctions and peacekeeping in Darfur, US interests in opposing the ICC over the years leading up to this decision were of such importance that the US prioritized this above other foreign policy interests and demonstrated willingness to veto resolutions to achieve its goals. As such, there was unusual political intensity within the Security Council over the decision to refer Darfur to the ICC. As described by an official involved in the negotiations, for the two or three months leading up to resolution 1953, it was 'the biggest resolution in town' (Author Interview 2011a). Another diplomat explained that he had not seen any other resolution involve such high-level diplomacy directly between state capitals (Author Interview 2011c).

Any interpretation of the ICC referral therefore needs to consider the unusual salience of this situation and strength of US opposition to the ICC. Given the strength of these national interests, there was less capacity for legitimation practices to influence the outcome than in some other situations, but even with intense national interests in play, external legitimation practices shaped the outcome of the negotiations, with the US prioritizing legitimation of itself as an actor over its unwillingness to legitimize the ICC via a referral.

The US had two overarching interests in this situation. One it achieved—negotiating exemptions for non-party states—and one it did not fully achieve—delegitimizing the ICC. Analysing whether and how the US was constrained in these negotiations will therefore depend on an assessment of these interests. The first of these—exemptions for non-party states—was clearly a high priority for the US. If this was the main priority for the US then one interpretation of this case is that the US position did not really change (Ralph 2009: 151). Before the Darfur negotiations, the US vetoed a peacekeeping resolution on Bosnia and Herzegovina and threatened to block peacekeeping decisions until non-party states were granted immunity, using its institutional power. The US had also attempted to negotiate bilateral non-surrender agreements with most states. During the Darfur negotiations, a US veto was widely thought to be possible until the day of the vote, and it was only after the US achieved a clause on non-party state immunity that it was willing to abstain and enable the referral. If this immunity was the primary strategic interest of the US, then it achieved this by threatening to veto the draft, and the US was not constrained (or only minimally constrained) by legitimation practices.

The second of the US' key interests—delegitimizing the ICC—was largely thwarted by the Darfur referral. An alternative interpretation of this case, then, is that the US legitimated the ICC by allowing the referral, despite its strong interest in not legitimating the Court (Ralph 2009: 151). Actions taken by the US indicate that this was also a high priority. The US had announced its intention not to ratify the Rome Statute. The bilateral non-surrender agreements can be seen as an attempt to delegitimize the ICC by circumventing the jurisdiction of the Court (Kelly 2007). During the Darfur negotiations, the US continued to take steps to delegitimize the ICC. Early in the negotiations, the US made it clear that it was not even willing to consider the ICC as an option for investigating crimes in Darfur, and as the US' ambassador-at-large for war crimes said, the reason it was not willing to consider this was because it did not want to legitimize the ICC. Statements made by US officials after the Darfur referral demonstrate that it maintained its fundamental opposition to the ICC and did not agree with the decision to refer the situation in Darfur. To achieve this, the US proposed an ad hoc tribunal and promoted it to the point of offering to pay for most of it. The US achieved some concessions, primarily an exemption for non-party states, but did not achieve its other key interest in delegitimizing the ICC as an institution via blocking the referral. Rather than fundamentally delegitimizing the ICC, the decision of the Security Council to refer its first case to the ICC is a significant endorsement of the ICC as an institution, cementing its role within international peace and security. In fact, the decision by the US to abstain on resolution 1593 can be seen as a shift in the US' interests on the ICC and a step towards a more constructive approach to the ICC (Birdsall 2010). The US was constrained in its attempts to delegitimize the ICC due to external legitimation practices, as it had been such a prominent and vocal advocate for Darfur that it was difficult to then veto a resolution aimed at ending impunity for a situation that the US had described as genocide. Legitimation practices—in this case legitimation of the US to domestic and international audiences—facilitated the US abstention and enabled the situation in Darfur to be referred to the ICC.

The advocate role that the US had adopted on Darfur in the UN Security Council meant that it was constrained in its opposition to the ICC referral. A US veto was thought to be a real possibility by those privy to the negotiations, until the day the resolution passed (Author Interview 2011a, 2011b, 2011c). If the referral had concerned a case in which the US had invested less diplomatic and political capital, it seems plausible that it would have vetoed a referral, given the US' long-standing resistance of the ICC. It took considerable diplomatic skill and competence by France and the UK to frame this as a binary choice for the US: continued impunity in Sudan or an ICC investigation (Bosco 2014: 109–110). The strong stance that the US had taken on Darfur both within the Security Council and domestically meant that it did not want to be seen to be aiding impunity, which is what a veto would have implied. US officials also thought that it was not worth jeopardizing relations with their European counterparts over the referral (Hamilton 2011: 58 and 65). This tension for the US facilitated the Security Council's first referral of a case to the ICC. However, the referral indicated tolerance towards the referral, rather than 'a political commitment to the prosecution of crimes by the ICC' (Aloisi 2013: 160). The US achieved some of its interests in these negotiations but was also constrained by the prominent stance that it had already taken in relation to Darfur.

This analysis has focused on the role of the US, since the negotiations reveal that the primary obstacle to this decision was US foreign policy. On the issue of sanctions in Chapter 4, China was the primary, but not only, opponent. China is also a non-signatory to the Rome Statute, so why did the negotiation hinge on stopping a US veto rather than preventing a Chinese veto? Various sources state that Khartoum was surprised that China didn't shield it with a veto (Taylor 2010: 181; Wenqi and Xinyu 2016: 96). One concession that was made to alleviate Chinese concerns over perceptions in this case was in relation to timing. Resolution 1593 passed on 31 March 2005, under the Brazilian presidency of the Security Council. Diplomats involved in the negotiations said that China was clear that it did not want the resolution tabled in April under the Chinese presidency, which would have posed greater difficulties for China to abstain and allow the resolution to pass (Author Interview 2011a, 2011c). China was not comfortable with the idea of symbolically endorsing the resolution under its presidency. There was also strong support for the referral from African states, which encouraged China to allow the resolution to pass (Author Interview 2011a). After the vote, China stated, 'We cannot accept any exercise of the ICC's jurisdiction against the will of non-State parties', concerns which led them to abstain, while also arguing that 'Undoubtedly, the perpetrators must be brought to justice' (United Nations 2005a: 5). Another reason that China chose not to veto could have been that it saw the high level of pressure brought to bear on the US for its possible veto.

At the time of writing, the ICC has not yet held anyone accountable for crimes committed in Darfur. Investigations have been instigated against six individuals, including charges of crimes against humanity, war crimes, and genocide, against the former President of Sudan, Omar al-Bashir. Of the six individuals, charges were not confirmed in one case, one person surrendered himself and is in custody, and the other four remain at large. In fact, as Kersten highlights, 'no arrest warrant following a Security Council referral has ever been enforced' (2016: 166). Omar al-Bashir remained President of Sudan for a decade after the ICC first issued a warrant for his arrest.

## Conclusion

The decision to refer the situation in Darfur to the ICC was unique among the range of decisions the Security Council has made on Darfur. Unlike decisions on sanctions and peacekeeping, where the focus of the negotiations was how best to address the situation in Darfur, for resolution 1593 the primary focus of the negotiations was on the tool for addressing impunity—the ICC—rather than on Darfur. However, the fact that the case was Darfur was what facilitated the US abstention. As the first referral of a situation to the ICC by the Security Council, it represented an initial negotiation of the relationship between these two institutions. As such, the Security Council referring a case to the ICC represented an endorsement of the ICC by the Council, which was especially significant given that three of the five permanent members—the US, China, and Russia—were not signatories to the Rome Statute. The referral of Darfur was also contentious because the Security Council demonstrates such a strong preference for 'agreed language', which meant that the language and framing used for this first referral was likely to be repeated as an internal legitimation practice, as it was in the Libyan referral (Dunne and Gifkins 2016; Gifkins 2016; Werner 2017). Given the intense opposition of the US to the ICC in the years leading up to the Darfur referral, this decision was also unusual in the level of national interest involved for a permanent member of the Security Council. The precedent-setting nature of this decision, combined with strong national interests of the US, meant that the referral was uniquely contentious among decisions on Darfur.

Given the strength of the national interests involved in these negotiations, some might expect that there would be no scope for legitimation practices to enable a particular outcome. Yet these negotiations show that, while the US obtained some concessions, it still did not want to legitimize the ICC with a referral, yet it was constrained from casting a veto vote. The legitimacy costs to the US for potentially vetoing the referral of Darfur to the ICC proved too high, since it had invested so much political capital in becoming a key advocate for Darfur. As Mills and Lee have framed it, the US 'lost its first test of wills over the ICC and was forced to stand back and allow the ICC to gain legitimacy' (2007: 511). The constraints on the US in this case came from legitimating itself as an actor and being perceived as legitimate by domestic audiences—an external legitimation practice—after having established itself as a key advocate over Darfur. The US wanted to maintain a consistent argument on Darfur, having established itself as a vocal advocate. Internal legitimation practices were also instrumental in achieving a decision because language which had been previously agreed in a resolution on Liberia was used to broker support and to enhance the legitimacy of the decision.

It was unknown until the day of the vote whether the US would abstain or veto. The US is not far behind Russia on the most use of the veto since the end of the Cold War, however, almost all US vetoes are on the Israel/Palestine conflict, so there is a higher political threshold for it to cast a veto on a different issue area. The US veto on peacekeeping in Bosnia and Herzegovina indicated that the ICC was of sufficient interest for the US to broaden its veto targets, however, it was difficult for the US to veto a draft resolution on Darfur specifically. If the US had chosen to veto the referral of Darfur to the ICC, this would have been difficult to align with the activist stance it had taken on Darfur both inside the Security Council and domestically. As Hurrell argued, 'Legitimacy implies a willingness to comply with rules or to accept a political order even if this goes against specific interests at specific times' (2005: 16). European states—particularly the UK and France—were determined to refer Darfur to the ICC, and since the US works collaboratively with them on most issues, it would have been damaging to their working relationships. A veto would also have been difficult for the US to justify domestically, where a strong civil society movement on Darfur had developed. Hamilton concluded that the civil society movement on Darfur did not play a significant role in championing the ICC specifically, but it did champion accountability more broadly, and generated a strong focus on Darfur in the US Congress and in public discourse. She argues that the diplomatic manoeuvring of European states and relationships between the US, UK, and France, were instrumental in facilitating a US abstention (Hamilton 2011: 68). The sequence of events was also bound up in US politics because it was the US declaration on genocide in Darfur which led to the US championing the establishment of the ICID in the Security

Council. Once the ICID recommended referral to the ICC, the momentum behind this issue put the US in a difficult position politically.

The referral of Darfur to the ICC—which hinged on the US choosing not to cast a negative vote—was enabled because of external legitimation practices for the US as an advocate on Darfur and internal legitimation practices of repeating agreed language. The prospective referral brought up core national interests for the US, as it had led a long-term campaign against the ICC, via unsigning the Rome Statute, creating bilateral non-surrender agreements, attempting to establish an alternative court, and vetoing a draft on Bosnia and Herzegovina. While the US achieved some of its goals—especially creating a carve out for non-signatory states—the referral still legitimized the ICC by solidifying its relationship with the Security Council. The US was clear before this resolution passed that it did not want to legitimize the ICC by referring a case to it, and went to extraordinary lengths to prevent this from happening. It maintained after the resolution passed that it had not changed its long-standing objections to the ICC. But when it came down to the wire, the US chose not to veto the draft because Darfur was a case where it had invested significant political capital, which made it very difficult to veto without losing credibility. The legitimation practices constrained the US from casting a veto, despite its strong interests in delegitimizing the ICC.

# Bibliography

Aloisi, Rosa. 2013. 'A Tale of Two Institutions: The United Nations Security Council and the International Criminal Court'. *International Criminal Law Review* 13: 147–168.

Arbour, Louise. 2014. 'The Relationship between the ICC and the UN Security Council'. *Global Governance* 20: 195–201.

Arieff, Irwin. 2005. 'Most UN Council Members Favor ICC Role in Darfur'. *Reuters*, 17 February.

Author Interview. 2011a. *Author Interview with an Official Involved in the Negotiations*. 9 September, Copenhagen.

Author Interview. 2011b. *Author Interview with Colin Keating from Security Council Report*. 17 February, New York.

Author Interview. 2011c. *Author Interview with P5 Diplomat*. 12 September, Paris.

Author Interview. 2011d. *Author Interview with Professor Edward C. Luck*. 14 February, New York.

Barnett, Michael N. 2002. *Eyewitness to a Genocide: The United Nations and Rwanda* New York: Cornell University Press.

Birdsall, Andrea. 2010. 'The "Monster That We Need to Slay"? Global Governance, the United States, and the International Criminal Court'. *Global Governance* 16(4): 451–469.

Bosco, David L. 2014. *Rough Justice: The International Criminal Court in a World of Power Politics*. Oxford: Oxford University Press.

Bravin, Jess and Scot J. Paltrow. 2005. 'Washington's Darfur Dilemma—Genocide Investigation Could Land in the International Criminal Court That Us Opposes'. *The Wall Street Journal*, 17 January.

Condorelli, Luigi and Annalisa Ciampi. 2005. 'Comments on the Security Council Referral of the Situation in Darfur to the ICC'. *Journal of International Criminal Justice* 3: 590–599.

Cryer, Robert. 2006. 'Sudan, Resolution 1593, and International Criminal Justice'. *Leiden Journal of International Law* 19(1): 195–222.

Dunne, Tim and Jess Gifkins. 2011. 'Libya and the State of Intervention'. *Australian Journal of International Affairs* 65(5): 515–529.

Dunne, Tim and Jess Gifkins. 2016. 'Libya and the State of Intervention (reprint)'. In *Civilian Protection in the Twenty-First Century*, eds. Cecilia Jacob and Alistair Cook. New Delhi: Oxford University Press.

Elagab, Omer Yousif. 2009. 'Indicting the Sudanese President by the ICC: Resolution 1593 Revisited'. *The International Journal of Human Rights* 13(5): 654–667.

Fehl, Caroline. 2009. 'Explaining the International Criminal Court: A Practice Test for Rationalists and Constructivist Approaches'. In *Governance, Order, and the International Criminal Court: Between realpolitik and a cosmopolitan court*, ed. Steven C. Roach. Oxford: Oxford University Press.

Gifkins, Jess. 2016. 'R2P in the UN Security Council: Darfur, Libya and Beyond'. *Cooperation and Conflict* 51(2): 148–165.

Global Policy Forum. 2005. 'Draft Security Council Resolution on Sudan'. Accessed 18 May 2017. Available at https://www.globalpolicy.org/component/content/article/206/39640.html.

Grünfeld, Fred, Wessel Vermeulen and Jasper Krommendijk. 2014. *Failure to Prevent Gross Human Rights Violations in Darfur: Warnings to and Responses by International Decision Makers (2003–2005)*. Leiden: Brill Nijhoff Publishers.

Hamilton, Rebecca. 2011. *Fighting for Darfur: Public Action and the Struggle to Stop Genocide*. New York: Palgrave Macmillan.

Heyder, Corrina. 2006. 'The UN Security Council's Referral of the Crimes in Darfur to the International Criminal Court in Light of the US Opposition to the Court: Implications for the International Criminal Court's Functions and Status'. *Berkeley Journal of International Law* 24(2): 650–671.

Hoge, Warren. 2005. 'France asking UN to refer Darfur to international court'. *New York Times*, 24 March.

Human Rights Watch. 2005. *EU Should Push for ICCR of Darfur during Rice Visit*. Available at https://www.hrw.org/news/2005/02/09/eu-should-push-icc-referral-darfur-during-rice-visit.

Hurrell, Andrew. 2005. 'Legitimacy and the Use of Force: Can the circle be squared?'. *Review of International Studies* 31(1): 15–32.

Jain, Neha. 2005. 'A Separate Law for Peacekeepers: The Clash between the Security Council and the International Criminal Court'. *The European Journal of International Law* 16(2): 239–254.

Johansen, Robert C. 2006. 'The Impact of US Policy toward the International Criminal Court on the Prevention of Genocide, War Crimes, and Crimes Against Humanity'. *Human Rights Quarterly* 28(2): 301–331.

Kelley, Judith. 2007. 'Who Keeps International Commitments and Why? The International Criminal Court and Bilateral Nonsurrender Agreements'. *American Political Science Review* 101(3): 573–589.

Kelley, Kevin. 2005. 'US Stand on ICC Will Delay Darfur Trials'. *All Africa*, 8 February.

Kersten, Mark. 2016. *Justice in Conflict: The Effects of the International Criminal Court's Interventions on Ending Wars and Building Peace*. Oxford: Oxford University Press.

Lederer, Edith M. 2004a. 'US Drops UN Bid for War Crimes Shield'. *Associated Press Newswires*, 23 June.

Lederer, Edith M. 2004b. 'US Faces Growing Opposition to Its Demand for Exemption of US Peacekeepers from War Crimes Prosecutions'. *Associated Press Newswires*, 22 May.

Leopold, Evelyn. 2004a. 'US Considers Compromise on Immunity for US Troop'. *Reuters*, 22 June.

Leopold, Evelyn. 2004b. 'US Lacks Votes for Immunity from War Crimes Court'. *Reuters*, 19 June.

Leopold, Evelyn. 2005a. 'Annan Asks Split Un Council to Move Fast on Darfur'. *Reuters*, 2 February.

Leopold, Evelyn. 2005b. 'Nigeria Wants New Panel on Sudan Crimes, Not ICC'. *Reuters*, 16 March.

Leopold, Evelyn. 2005c. 'US Lobbies United Nations for New Darfur Court'. *Reuters*, 28 January.

Leopold, Evelyn and Arshad Mohammed. 2005. 'Update 2—US Might Permit Icc Trials for Sudan Suspects'. *Reuters*, 31 March.

Lynch, Colum. 2004. 'China May Veto Resolution on Criminal Court: Beijing Says UN Motion Could Shield US Troops from Abuse Allegations'. *The Washington Post*, 29 May.

Mearsheimer, John T. 1994/1995. 'The False Promise of International Institutions'. *International Security* 19(3): 5–49.

Mills, Kurt and Anthony Lee. 2007. 'From Rome to Darfur: Norms and Interests in US Policy toward the International Criminal Court'. *Journal of Human Rights* 6(4): 497–521.

Ralph, Jason. 2007. *Defending the Society of States: Why America Opposes the International Criminal Court and Its Vision of World Society*. Oxford: Oxford University Press.

Ralph, Jason. 2009. 'Anarchy Is What Criminal Lawyers and Other Actors Make of It: International Criminal Justice as an Institution of International and World Society'. In *Governance, Order, and the International Criminal Court: Between Realpolitik and a Cosmopolitan Court*, ed. Steven C. Roach. Oxford: Oxford University Press.

Rome Statute of the International Criminal Court. 1998. Available at http://legal.un.org/icc/statute/english/rome_statute(e).pdf.

Schabas, William A. 2004. 'United States Hostility to the International Criminal Court: It's All about the Security Council'. *European Journal of International Law* 15(4): 701–720.

Schabas, William A. 2010. 'The International Criminal Court'. In *The International Politics of Mass Atrocities: The Case of Darfur*, eds. David R. Black and Paul D. Williams. London and New York: Routledge.

Scheffer, David J. 2001. 'Staying the Course with the International Criminal Court'. *Cornell International Law Journal* 35(1): 47–100.

Stedjan, Scott and Colin Thomas-Jensen. 2010. 'The United States'. In *The International Politics of Mass Atrocities: The Case of Darfur*, eds. Paul D. Williams and David R. Black. London and New York: Routledge.

Taylor, Ian. 2010. 'The People's Republic of China'. In *The International Politics of Mass Atrocities: The Case of Darfur,* eds. Paul D. Williams and David R. Black. London and New York: Routledge.

United Nations. 2002a. *4563rd Meeting United Nations Security Council.* New York, 30 June, S/PV.4563.

United Nations. 2002b. *4568th Meeting United Nations Security Council.* New York, 10 July, S/PV.4568.

United Nations. 2002c. *Security Council Resolution 1422.* New York, 12 July, S/RES/1422.

United Nations. 2003a. *4772nd Meeting United Nations Security Council.* New York, 12 June, S/PV.4772.

United Nations. 2003b. *4803rd Meeting United Nations Security Council.* New York, 1 August, S/PV.4803.

United Nations. 2003c. *Security Council Resolution 1497.* New York, 1 August, S/RES/1497.

United Nations. 2004. *Resolution 1564.* 18 September, S/RES/1564.

United Nations. 2005a. *5158th Meeting of the United Nations Security Council.* New York, 31 March, S/PV.5158.

United Nations. 2005b. *Report of the International Commission of Inquiry on Darfur to the United Nations Secretary-General.* Geneva, 25 January. Available at http://www.un.org/news/dh/sudan/com_inq_darfur.pdf.

United Nations. 2011. *Security Council Resolution 1970.* New York, 26 February, S/RES/1970.

US Department of State. 2005a. *International Community Must Act On Darfur, Says Rice—Secretary, Hungarian Foreign Minister Discuss Ukraine, Iraq, Balkans.* 1 April.

US Department of State. 2005b. *US Decision on UN Darfur Resolution Difficult, Burns Says—Under Secretary of State Discusses UN Vote, Situation in Sudan.* 1 April.

US Embassy Dar es Salaam. 2005. *Tanzanians Cautious on AU-UN Sudan Tribunal.* Wikileaks. Available at https://wikileaks.org/plusd/cables/05DARESSALAAM206_a.html.

US Embassy Paris. 2005. *Darfur Accountability: Amb. Prosper's Discussion with French MFA.* Wikileaks. Available at https://wikileaks.org/plusd/cables/05PARIS1039_a.html.

Wadhams, Nick. 2005. 'France Proposes Resolution Referring Sudan to International Criminal Court, Forcing US into a Dilemma'. *Associated Press Newswires,* 23 March.

Wenqi, Zhu and Leng Xinyu. 2016. 'China in the Security Council'. In *The UN Security Council in the Twenty-First Century,* eds. Sebastian von Einsiedel, David M. Malone, and Bruno Stagno Ugarte. Boulder: Lynne Rienner.

Werner, Wouter. 2017. 'Recall It Again, Sam. Practices of Repetition in the Security Council'. *Nordic Journal of International Law* 86(2): 151–169.

# 7

# Peacekeeping and the Challenge of Consent

*I was never a fan of that mission … it was not born under the right star.*
Former Under-Secretary-General for Peacekeeping Affairs,
Jean-Marie Guéhenno referring to the AU/UN hybrid
peacekeeping operation in Darfur (cited in Lynch 2014b)

The African Union/United Nations hybrid peacekeeping operation in Darfur
(UNAMID) was authorized more than four years into the conflict in Darfur and
was contentious from the start. This chapter analyses the 'birth' of UNAMID,
growing out of a process of failure and compromise within the Security Council. Prior to UNAMID, from 2004 to 2007 the African Union Mission in Sudan
(AMIS) was in Darfur but struggled with lack of finances and equipment. The
Western states funding AMIS wanted to shift to UN peacekeeping—where costs
are shared between UN members—which aligned with activism from civil society
groups calling for more effective protection of civilians in Darfur. Replacing AMIS,
however, proved to be a herculean feat, which took two years of negotiations
within the Security Council and exposed real limits in the Council's authority.
Security Council members initially worked from the erroneous assumption that
Khartoum would consent to UN peacekeeping and authorized a transfer from
AMIS to the United Nations Mission in Sudan (UNMIS).[1] This assumption led to
a legally binding resolution which could not be implemented—because UN peacekeeping is premised on the consent of the host state. As such, resolution 1706 has
been described by Jean-Marie Guéhenno as the resolution that 'may have been
the Council at its worst' in terms of the gap between words and deeds (Author
Interview 2011b). Following from this failure, it took Security Council members
another year to negotiate and authorize a compromise proposal of hybrid AU/UN
peacekeeping in the form of UNAMID.

Key actions and turning points in these negotiations were driven by external legitimation practices. Throughout the first stage—where Western drafters
attempted to transfer from AU peacekeeping to UN peacekeeping—drafters
prioritized 'doing something' over doing something that could be implemented.
Khartoum regularly stated during this period that it would not consent to UN

[1] UNMIS was already operating in the southern region of Sudan at the time (now South Sudan).

*Inside the UN Security Council.* Jess Gifkins, Oxford University Press. © Jess Gifkins (2023).
DOI: 10.1093/oso/9780192869029.003.0008

peacekeeping, and indeed, that it was actively hostile to the idea, yet drafters proceeded on the basis of 'doing something' or being seen to be doing something and authorized UN peacekeeping anyway. Then following the failure of resolution 1706; drafters recommitted to their usual internal legitimation practice of prioritizing unanimity, and sought to obtain and maintain Khartoum's consent, realizing that consent and unanimity were essential for the legitimacy of a decision and for the capacity to implement the decision. Following wars in Iraq and Afghanistan, Western states were not interested in pursuing non-consensual military action in Darfur (Booth Walling 2013: 210). Alongside Western states' realization that consent was critical, China's position shifted towards actively encouraging Khartoum's consent. China sought the social status of legitimacy and worked constructively with Western drafters in these negotiations to secure Khartoum's consent and to bring other Security Council members on board. Western states' attempts to 'do something' and China's constructive actions to secure Khartoum's consent can both be understood as external legitimation practices. Security Council members' drive for legitimation shaped these key actions and turning points.

The other key legitimation practices evident in these negotiations were the influence of the UN Secretariat—drawing from its legitimated authority—and the way that drafters sought support from African states to bolster the legitimacy of the drafts. Alongside the actions of Security Council members, actions taken by the Secretariat were also instrumental in shaping outcomes. The mandates set out in both resolution 1706 and resolution 1769 were recommended to the Security Council by the Department of Peacekeeping Operations (DPKO),[2] which gave it influence over the mandate, but also exposed the limits to its influence as the political concerns expressed about peacekeeping in a hostile environment were not heeded. The eventual breakthrough in the form of a novel 'hybrid' peacekeeping operation came from the office of the UN Secretary-General, which designed the plan and lobbied Security Council members and Khartoum to accept it. This represented considerable influence for the Secretariat, which went beyond technical advice to create a plan and set out a strategy to bring Khartoum on board. The final way that legitimation shaped negotiations was drafters seeking support from regionally relevant states. Drafters sought the support of African states—knowing that the support of key African states would make it harder for other states to dissent, as an internal legitimation practice. Where von Billerbeck (2019, 2020) demonstrates the self-legitimation of international organizations, including the UN's Department of Peacekeeping Operations, by showing how the institution is legitimated to and by staff, the focus of this analysis is on the legitimation of states involved in decision-making, and decisions themselves. The legitimacy

---

[2] The department has since been renamed the Department of Peace Operations; however, this chapter refers to it under the name it had at the time since that it what it was called in quotes and interviews.

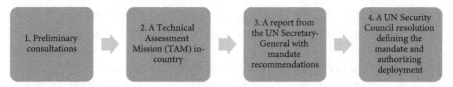

**Figure 7**  Authorizing UN Peacekeeping Operations

*Note*: Adapted from DPKO 2008 (47–49). In the case of Darfur, this cycle was conducted twice. In the second cycle, the Technical Assessment Mission was replaced with a Quick Review Mission to update the earlier Technical Assessment Mission.

of the UN Secretariat and the legitimacy of regionally relevant states was evident in the ways that their input was sought or valorized.

This chapter proceeds in three sections. The first section provides contextual background on three key areas: the evolution of China's stance on peacekeeping and consent, the drivers towards UN peacekeeping in Darfur, and the Darfur Peace Agreement. The second section charts the process towards failed resolution 1706 in stages with a Technical Assessment Mission, a report from the UN Secretary-General, and then drafting the resolution. The third section charts the process towards resolution 1769, which established UNAMID, by outlining how the impasse left by the failure of resolution 1706 was breeched, planning the hybrid operation, and drafting resolution 1769. Peacekeeping operations are mandated via a sequential process, with four key stages (DPKO 2008: 47–49). Each of these stages occurred twice in this case: once towards UN peacekeeping, and then again towards AU/UN peacekeeping (see Figure 7). These stages are premised on a viable peace agreement and there being a 'peace to keep' (DPKO 2008: 49–50), which was not the case in Darfur.

## Context for Negotiations on UN Peacekeeping in Darfur

### China and Peacekeeping

China's stance on peacekeeping in Darfur represents an evolution in Beijing's attitude towards peacekeeping, as it was willing to actively seek Sudan's consent for this operation. Historically, China has been wary about peacekeeping as an incursion on sovereignty. Through the 1990s, and especially since 2000, China has become more willing to authorize UN peacekeeping and more involved in peacekeeping operations (Hirono and Lanteigne 2011; Stähle 2008).[3] China retains two fundamental priorities for peacekeeping: UN authorization, and consent from the host state, both of which are in line with the Security Council's usual practices

---

[3] At the time of writing, China was the highest troop-contributing country out of the P5, contributing the tenth highest number of troops of all countries.

(Lanteigne 2011).[4] Consent poses a significant dilemma, however, when it is the state itself that is responsible for the violence (Orchard 2010). Through the 1990s, China regularly abstained on Security Council resolutions that authorized the use of force or issued heavily caveated statements on why a specific operation should be allowed, despite its general opposition to the use of force. However, for peace-keeping in East Timor, China voted affirmatively without issuing a statement at the vote (Stähle 2008). Beijing supported two operations in East Timor in 1999, despite Indonesia having been coerced to consent to these missions (Carlson 2004; Jago 2010). China's support for operations in East Timor can be seen as a turning point in Beijing's interpretation of host-state consent, as consent was not volun-tarily given by Indonesia yet China supported the resolutions (Lanteigne 2011). China's role in relation to peacekeeping in Darfur represents a further evolu-tion in China's primacy on host-state consent. Khartoum did, eventually, consent to hybrid AU/UN peacekeeping in Darfur, and the key reason for this consent was pressure and encouragement directly from Beijing. This was remarkable, as Holslag demonstrates; it was the first time 'China actively sought to persuade a sovereign government to assent to the deployment of blue helmets on its national soil' (2008: 83). Not only was China willing to accept coerced consent for UN peacekeeping, but it was also willing to participate in securing this consent itself.

It has been argued elsewhere that China's support for peacekeeping in Darfur was an effect of the 'Genocide Olympics' campaign (see for example Haeri 2008: 39; Lanz 2009: 672). The term 'Genocide Olympics' had been used previously, but it was popularized in an op-ed in March 2007 by Mia Farrow and her son Ronan Farrow (Hamilton 2011: 139). This movement has been credited with some spe-cific successes, such as China's creation of a special envoy on Darfur (for discussion see Hamilton 2011: 140). However, the most significant actions taken by Beijing to secure the consent of Khartoum and enable a unanimous resolution on UNAMID predate the 'Genocide Olympics' movement. As others have argued, the timeline doesn't add up for the 'Genocide Olympics' movement to be the primary driver of China's support for UN peacekeeping in Darfur (Fung 2019). Significantly, in November 2006, Beijing was actively involved in encouraging Sudan's consent for peacekeeping.

How, then, do we explain China's advocacy with Khartoum over peacekeep-ing? During the negotiations towards UN peacekeeping in early 2006, China was only willing to join other states in calling for Sudanese consent (Holslag 2008: 77). Yet by late 2006, China was engaged in bilateral diplomacy with Khartoum to encourage its consent. There are two different explanations for this change.[5] First,

---

[4] An additional growing concern for China is regional support (for discussion see Lee, Chan, and Chan 2012).

[5] A third possible explanation is economic interests (see, for example, Large 2008), however, while China has economic interests in Sudan, it is difficult to account for a shift in China's actions in relation to ongoing economic interests in Sudan.

the impact of bilateral diplomacy by both the US and African states contributed towards a shift from China. A Chinese diplomat at the UN described the impact of advocacy from the US president to the Chinese president in May 2006 and said that this had encouraged increased support from China within in the Security Council (US Permanent Mission to the UN 2006b). Also, in early November 2006, Beijing hosted the Forum on China–Africa cooperation. During this event, many African states expressed concerns to China over Darfur (Wenqi and Xinyu 2016: 97) and Chinese president Hu Jintao publicly expressed the need for a settlement of the Darfur conflict and said that there was a role for China in contributing to this (Xinhua 2006). Second, the reputational costs for China were particularly high, given that it was aiming to establish itself as a 'responsible power' (Richardson 2011; Wuthnow 2010). In seeking legitimation for itself as an actor there were incentives for China to shift its position towards Western norms. China has moved closer to dominant norms in the Security Council over time, with increased support for peacekeeping (Kent 2002; Morphet 2000; Wenqi and Xinyu 2016). China has also demonstrated unwillingness to block decisions alone in the Security Council. At the time of writing, China had not cast a lone veto since the 1990s and has made 'average' use of its veto since the end of the Cold War, having cast fewer vetoes than the US and Russia and more than the UK and France.[6]

These two explanations are not necessarily separable; bilateral diplomacy could have encouraged China to show greater flexibility towards Western and African positions for reputational gain, and the desire for reputational gain could have made China more receptive to bilateral diplomacy. Fung sets out the case that China's shift to encourage consent from Khartoum was due to Beijing's status sensitivity whereby both its key peer groups—great powers and the Global South (in this instance represented by African states)—encouraged China to support UN peacekeeping (Fung 2019: 64). This status sensitivity was connected to the legitimation of Beijing in the eyes of key audiences. Diplomats from Western states, African states, and the AU all presumed that China had leverage over Khartoum, and encouraged China to use this leverage. It was an external legitimation practice—legitimacy in the eyes of international audiences—that led to the shift whereby China lobbied Khartoum for consent.

## The Drive for UN Peacekeeping

The African Union's AMIS peacekeeping operation struggled for sufficient resources to provide adequate protection. AMIS began in Darfur with the deployment of a small group of observers to monitor the N'Djamena Humanitarian Ceasefire Agreement in April 2004 (African Union 2004). From modest initial

---

[6] Data compiled by the author from UN (2022).

goals, AMIS expanded to a force of over 7,000 and a civilian protection mandate. The operation was plagued by problems, however, with a small number of troops responsible for an area the size of France, differing interpretations of their mandate, and lack of resources, both financial and logistical (Badescu and Bergholm 2010: 103–107). Furthermore, the government of Sudan engaged in persistent obstruction of AMIS on the ground (Duursma 2021: 14). As explained by key author on Sudan Alex de Waal, 'The standard formula for describing AMIS was that it was overstretched, ill-equipped and undermandated' (2007: 1045). Despite early successes, AMIS struggled due to small numbers of troops—who were sometimes unpaid—operating with a lack of equipment and an uncertain mandate (Badescu and Bergholm 2010: 103–107; Flint and de Waal 2008: 176–179). However, there was no appetite from member states to lead a non-consensual peace enforcement operation, which meant that consent-based peacekeeping was the only viable option (de Waal 2007: 1047; Flint and de Waal 2008: 268; Hamilton 2011: 112). Notably, the drive towards UN peacekeeping occurred well after the height of violent conflict in Darfur in 2003 and early 2004.[7] Only by 2006 was the momentum for UN peacekeeping sufficient to deploy troops, yet this was disconnected with the latest situation on the ground, where the primary cause of death was disease rather than violence, and there was no 'peace to keep'.

The drive towards UN peacekeeping in Darfur began in 2006 with the alignment of two motivations. First, and most importantly, donors did not want to continue to fund the struggling African Union peacekeeping operation. The main countries who were funding AMIS—the US, the UK, Norway, the Netherlands, France, and Canada—were frustrated by its lack of effectiveness and were reluctant to continue funding (Hamilton 2011: 79; MacKinnon 2010: 91–92). Approaches to funding were key here: while AMIS was funded voluntarily by a small group of states, UN peacekeeping operations are funded via assessed contributions from all members of the United Nations, so the costs are shared more widely. The funding imperative created the political impetus for a shift from AU peacekeeping to UN peacekeeping.

Second, UN peacekeeping was expected to be more effective at civilian protection than the struggling AU force, and this aligned with US civil society activism on Darfur.[8] Activism on Darfur—under the umbrella of the Save Darfur Coalition— had grown dramatically by 2006 and has been compared in size and scale with the movements against the Vietnam War and against apartheid in South Africa (Flint and de Waal 2008: 183; Mamdani 2009: 22).[9] The Save Darfur Coalition aimed

---

[7] For further discussion on this point see Chapter 3.

[8] Rebecca Hamilton's 'Fighting for Darfur' (2011) is an excellent resource on US civil society and Darfur, as is David Lanz's (2020) 'The Responsibility to Protect in Darfur'.

[9] Other conflicts at the time such as in Angola and in the Democratic Republic of the Congo had higher mortality rates, but it was the conflict in Darfur that gained wider public and political attention see Mamdani (2009: 21).

to influence US foreign policy on Darfur and ran powerful campaigns calling for expanded peacekeeping, such as coordinating one million people to send post-cards to US Congress officials (Hamilton 2011: 80–82). To be clear, the primary motivation for UN peacekeeping was not to address a deteriorating security situa-tion in Darfur since security was not deteriorating at the time. By 2006, mortality rates in Darfur were below what is considered 'emergency level' and were compa-rable with pre-war mortality rates (Degomme and Guha-Sapir 2010; Flint and de Waal 2008: 173). The civil society mantra that 'things are getting worse' through 2005 to 2007 did not match reality on the ground, but it influenced public per-ceptions nonetheless (Flint and de Waal 2008: 187–191). By 2006, there came a somewhat unlikely convergence of interests between Western funders of AMIS and US domestic civil society, which was advocating for the transition from AU peacekeeping to UN peacekeeping. These converging interests drove the politics within the UN Security Council and led to a concerted push for UN peacekeeping.

## The Darfur Peace Agreement

As with previous negotiations on sanctions, the pressure towards UN peacekeep-ing was driven by the external legitimation practice of 'doing something', or to be seen to be doing something, along with the financial imperative. A precursor to establishing UN peacekeeping in Darfur was that there was a 'peace to keep'. Usu-ally, peacekeeping supports the priorities set out in a peace agreement, but in the case of Darfur this process was inverted: the peace agreement was driven by the international actors' desire for UN peacekeeping (de Waal 2007). This logic was stated simply by Michael Gerson, the chief speechwriter for US President Bush: 'the Darfur Peace Agreement was important for one reason—it paved the way to a UN mandated force' (cited in Hamilton 2011: 98). Peace talks on Darfur were pro-tracted, and by March 2006 they had reached their seventh round and the patience of donors had been exhausted (Nathan 2007: 248).[10] Attempting to conclude the peace talks, AU mediators issued a communiqué which set the end of April as the final deadline for a peace agreement. This deadline was endorsed by the UN Security Council in resolution 1663 in March, which was unanimous, and in a presidential statement in April to stress the seriousness of the deadline (United Nations 2006d: 1, 2006g: 1).

When this 'final deadline' was exceeded, US Deputy Secretary of State Robert Zoellick and UK Secretary of State for International Development Hilary Benn flew to Abuja on 2 May to extract an agreement (Toga 2007: 241). Under intense international pressure, the Darfur Peace Agreement (DPA) was signed on 5 May 2006. It was only signed by Khartoum and one of the rebel leaders, Minni Minawi,

---

[10] For a detailed account of the Darfur peace process, see de Waal (2007).

from the largest faction of the Sudan Liberation Movement/Sudan Liberation Army (SLM/A), while the Justice and Equality Movement (JEM) and smaller faction of the SLM/A led by Abdelwahid al-Nur refused to sign. With support from only one rebel group, and a lack of serious commitment from Khartoum, it was never a stable peace agreement. Not only was the resulting agreement unstable, but the DPA led to further splintering of rebel groups, which made the conflict harder to resolve (Nathan 2007). The drive for UN peacekeeping in Darfur led to a flawed peace agreement which exacerbated the conflict, yet this agreement formed the foundation of negotiations towards UN peacekeeping. The strategy was not driven by the situation on the ground but by financial pressures and the desire to 'do something', a frame that is also discussed by Lanz (2014, 2020).

Consent from the government of Sudan—a prerequisite for UN peacekeeping—became the primary obstacles in the drive towards peacekeeping. It was assumed not only that a peace agreement would create conditions for peacekeeping (i.e., 'a peace to keep'), but that it would also lead to Sudanese consent for peacekeeping. The US government had been told by Sudanese Vice President Taha that once a peace agreement was signed, Khartoum would consent to UN peacekeeping in Darfur. This understanding was passed on to the mediation team and consent was assumed—rather than discussed—in the peace talks (Hamilton 2011: 96–97; Toga 2007: 239). The assumption of Sudanese consent for UN peacekeeping proved flawed and led to the downfall of the transition to UN peacekeeping.

## Towards Resolution 1706

### Authorizing a Technical Assessment Mission

China was wary of peacekeeping in Darfur from the start, given the lack of consent from Khartoum, however, this wariness did not translate into outright obstruction on peacekeeping within the Security Council. Far from being obstructionist, the process of negotiating resolutions on peacekeeping shows Beijing's active involvement in encouraging Sudanese consent, and active engagement within Security Council negotiations to (at times) facilitate unity in decisions. The first step in creating a peacekeeping operation is deploying a Technical Assessment Mission (TAM), which would usually be just that: technical. In the case of Darfur, however, the government of Sudan had already expressed a lack of willingness to allow a TAM to proceed, so the assessment mission was authorized via Security Council resolution 1679—under Chapter VII—to make it legally binding. Significantly, there is evidence to suggest that the unity achieved in this resolution was made possible by actions of China within the negotiations. China sought to legitimize itself by assisting the US in passing a unanimous resolution.

During the negotiations towards resolution 1679, China, Russia, and Qatar were all expected to abstain. Unity was made possible, however, after China approached the US drafters to seek a unanimous vote for this resolution. China told the US that if the reference to resolution 1591 (on sanctions, on which China had abstained) was removed then Beijing would support the resolution (US Permanent Mission to the UN 2006b).[11] The cable states that the reference was removed at China's request, and not only did China vote affirmatively, but it also successfully encouraged Russia and Qatar to also vote affirmatively, enabling a unanimous vote (US Permanent Mission to the UN 2006b).[12] The Chinese UN delegation explained to the US UN delegation that bilateral diplomacy from the US was crucial in gaining Chinese support for this initiative. Notwithstanding the explicit recognition that this explanation 'clearly played to its audience', the cable describes interactions between the US president and the Chinese president as well as between US Deputy Secretary of State Zoellick and the Chinese Ambassador to the US (US Permanent Mission to the UN 2006b). In these interactions, US officials stressed to Chinese officials how important UN peacekeeping in Darfur was to the US.[13] These explanations are also congruent with a US cable the following month, which confirmed that the Chinese mission approached the US mission to request US 'partnership' in encouraging Khartoum to agree to the transition to UN peacekeeping (US Permanent Mission to the UN 2006a). A Chinese UN diplomat reported to the US that China had already succeeded in alleviating some concerns that Khartoum held about the transition (US Permanent Mission to the UN 2006a). The unanimous vote for resolution 1679 was made possible by China's activism in seeking unity for this decision. This shift in China's position was attributed by Chinese diplomats to bilateral diplomacy. There were also reputational and legitimation benefits for China in assisting the US in achieving its goals.

The other key impact of legitimation practices on this negotiation was the influence and instrumentalization of support from the AU and from African states. In addition to authorizing a TAM, the resolution included a threat of targeted sanctions: 'such as a travel ban and assets freeze, against any individual or group

---

[11] This comes from a cable by the US delegation at the UN which was released by WikiLeaks. The cable is signed by US Ambassador John Bolton, however, the signature on cables does not necessarily indicate that the person wrote or even saw the cable. For discussion see Keating (2010).

[12] The final version of Resolution 1679 contains a reference to resolution 1591 in the preamble, but we can infer from the cable and the final text that there was a further reference to resolution 1591 that China resisted was removed.

[13] Since these conversations and negotiations were all off-the-record, there are limited sources to corroborate this version of events. An advantage of this source is that it is an account of the negotiations from the time the negotiations occurred. It is also corroborated by China's similar actions in June 2006 and November 2006 to seek Sudanese consent for UN peacekeeping, and to request US partnership in encouraging consent (see respectively: US Permanent Mission to the UN 2006a; Brookings Institute 2006). An alternative account is presented in John Bolton's memoirs, in which he says that the US willingness to put the draft to a vote whether there was unity or not led to dissenters backing down and voting affirmatively, however, this does not match the account recorded by US diplomats at the time (2007: 353–354).

that violates or attempts to block the implementation of the Darfur Peace Agreement' (United Nations 2006e: 2). The resolution specifies that this provision was included in response to a request by the African Union. The request from a regional group was significant: resolution 1679 was the first resolution on Darfur that included a provision on sanctions and secured unanimous support. The three states that had concerns over this resolution—China, Russia, and Qatar—each made statements after their affirmative vote. All three stated that AU support for the resolution was key to their support for the resolution (United Nations 2006a). There is also evidence that drafters sought African support for the resolution as an internal legitimation practice. A US cable from the time reported: 'We "co-chaired" negotiations with the Congolese delegation during our discussions of [UN Security Council resolution] 1679 (2006), strengthening our position with the more intransigent Members who were not as willing to contradict an African delegation on Sudan' (US Permanent Mission to the UN 2006f). As such, support from the AU and from African states was instrumental because of the legitimacy that regional support bestowed on the resolution—and can be seen as an internal legitimation practice—which helped to generate wider Council support.

## Drafting the UN Secretary-General's Report

The UN Secretariat provided advice to the Security Council on establishing a mandate for peacekeeping in Darfur, some of which was taken up and some ignored, demonstrating both the key influence of the Secretariat on the Council and the limits of this influence. Drawing on the information obtained by the TAM, the next stage of decision-making was a report from the UN Secretary-General outlining mandate options for the Security Council. The Report of the Secretary-General on Darfur was written primarily by DPKO, and it shaped the mandate for UNMIS' expanded role into Darfur. Although it is up to the Security Council to determine the mandate for peacekeeping operations, it usually follows the Secretariat's recommendations closely (Holt, Taylor and Kelly 2009: 119).[14] In this role, the Secretariat is seen as an expert and as an impartial information provider (Abbott and Snidal 1998; Barnett and Finnemore 2004). The Secretary-General's report was based on facts on the ground from UNMIS and the TAM and was drafted by DPKO's Darfur Planning Team (Author Interview 2011c). DPKO's role in drafting these reports is to take the information from the field and generate recommendations to the Security Council.

---

[14] For an example where the Security Council did not use DPKO's recommendations, see (Dijkstra 2014) on South Sudan. Dijkstra argues that, in that case, the US circumvented DPKO's guidance, as it had used its own shadow bureaucracy to conduct its own assessment into the requirements.

This advisory role gives DPKO considerable influence over peacekeeping mandates. The report gave three different 'options' for expanding UNMIS into Darfur, each of which had a different balance between troops and air power, however, the report makes clear which the Council 'should' choose (United Nations 2006c: 17–18). Option one was a force size of 17,300 troops, three fixed-wing aircraft, and up to twenty-six helicopters. This option is described in the report as 'optimal' and as the 'fastest route to a secure environment' (United Nations 2006c: 18). Option two was a force of 18,600 troops, but only thirteen helicopters, which the report advised would be the most challenging to deploy and to sustain. Option three was 15,300 troops and thirty-two helicopters, which the report advised would be slightly faster to deploy and easier to sustain but may not be able to protect civilians and would be vulnerable to weather constraints. Although these were presented as 'options' to the Council, the Secretariat clearly framed two of the three as poor options, both logistically and for civilian protection. Unsurprisingly, given the way these options were presented, the Council endorsed the troop level recommended in option one (United Nations 2006f: 3). The Security Council relies on the DPKO, as independent technical advisors, to make recommendations on force size and logistics, and typically adopts these recommendations wholesale. By presenting three options, DPKO could guide the Security Council towards its preferred option of middle numbers of troops and helicopters and had direct input into the authorized mandate, demonstrating considerable influence as a function of its legitimated authority.

On force size and logistics, DPKO had direct influence on the Security Council's resolution, however, this also represented the limits of its influence. DPKO has its own interests in limiting peacekeeping to situations with a reasonable likelihood of success (Allen and Yuen 2013; Dijkstra 2014). Staff within DPKO thought that sending peacekeepers to Darfur was imprudent, given the lack of consent and the instability of the DPA. The TAM reported that Khartoum unequivocally rejected a transition to UN peacekeeping (Holt, Taylor and Kelly 2009: 117). Likewise, Jean-Marie Guéhenno, then head of DPKO, briefed the Security Council and described deployment as 'a logistical challenge—more like a nightmare' given the size of Darfur and the levels of insecurity (US Embassy Ndjamena 2006). Other DPKO staff said that the conditions for peacekeeping were simply not present, but they made recommendations to the Security Council 'because we were being pressured to plan' (Hamilton 2011: 111). These pressures meant that DPKO drafted a report—as requested—which started from the assumption that peacekeeping was viable despite DPKO having real concerns, which shows the limits to its influence.

This is evident in the framing of the Darfur Peace Agreement within the Secretary-General's report. Even while this report was being drafted, UNMIS personnel advised the Secretariat that the DPA was not a viable peace agreement (Holt, Taylor and Kelly 2009: 118 and 348). Indeed, a DPKO staff member described the DPA as 'a meaningless agreement' (Author Interview 2011c).

Peacekeeping is premised on a 'peace to keep', however, so the DPA was foundational to the deployment of peacekeepers. These tensions are clear in the report, which included the phrase 'Darfur Peace Agreement' fifty times and treated the DPA as foundational for the expansion of UNMIS into Darfur, yet the final section of the report expressed significant concerns about the DPA. The final section highlighted that some rebel groups did not sign the DPA, cited lack of support from the Darfuri community, and suggested there were structural difficulties within the DPA (United Nations 2006c). Political pressures on DPKO meant that the Secretary-General's report was written from the assumption that the DPA was viable and that there was a 'peace to keep', despite the DPKO holding real concerns about this approach.

This seemingly inconsistent approach can be understood by looking at how the report was drafted. An individual from DPKO explained that most of the report was populated by information that came from UNMIS and from the TAM (Author Interview 2011c). However, when writing the final 'Observations and recommendations' section, DPKO had liberty to interpret the facts and give a more political perspective (Author Interview 2011c). DPKO used this small leeway to express real concerns about the viability of the DPA, however, these concerns went unheeded. Members of the UN Secretariat are at times under substantial political pressure, which shapes the analysis they provide, and it has been suggested that the inconsistencies in the July 2006 report are an example of this (Holt, Taylor and Kelly 2009: 102). This represented the limits of DPKO's influence. DPKO influenced the eventual mandate of UNMIS' role in Darfur, through its recommendations in the report, but its overall concerns about the viability of the DPA and that the mission would be a 'nightmare' were not heeded by the Security Council. The Secretariat holds legitimated authority and technical expertise, which allows it to provide recommendations, however, these recommendations interact with political priorities in the Security Council.


## Failed Attempt to Transfer to UN Peacekeeping: Resolution 1706

Despite the DPKO's concerns, Security Council members—led by the US—moved on to the next stage of the authorization process, which was drafting a resolution to authorize peacekeeping. Negotiations towards resolution 1706 were intense throughout the month of August 2006. The question of consent featured prominently in these negotiations, as it was unusual to authorize a peacekeeping operation without prior consent.[15] The other divisive issue was whether to include

---

[15] Although rare, another example is resolution 2303, which authorized deployment of police officers in Burundi without consent. For discussion see Jacobsen and Engell (2018).

language on the responsibility to protect (R2P), which by this time had been formalized by the UN General Assembly and endorsed by the UN Security Council but had not yet been applied to a country-specific conflict. Consent and R2P were the focus of debates throughout August; however, Khartoum was consistent in its message that it did not consent to UN peacekeeping. What this meant was that Council members found language that they could accept, rather than finding agreement on the substance of the decision. These text-based compromises facilitated resolution 1706, but it could not be implemented, due to the lack of consent from Khartoum. As such it exposed the limits to the Council's power—with a legally binding resolution that it could not implement—and was a threat to the Council's credibility (Hamilton 2011: 127; Lynch 2010; Williams 2010: 209). The political priority to change the funding arrangements, combined with the external legitimation practice of 'doing something' to assist with civilian protection, meant that passing a resolution was prioritized over passing a resolution which could be implemented.

There was a fundamental flaw within these negotiations: while Western states wanted UN peacekeeping in Darfur, UN peacekeeping is premised on consent of the host state, and the government of Sudan did not consent. Host state consent is one of the fundamental principles of peacekeeping for the DPKO (2008: 31).[16] DPKO also recognizes the complexity of consent, where it can be fluid and contingent, where formal consent can be revoked, and where formal consent does not necessarily equate to consent at the operational level (for discussion see DPKO 2008: 31–33). In contrast to this, US drafters relied on an overly simplistic understanding of consent, as an operational hurdle rather than as a fundamental prerequisite. In the absence of consent, the US wanted a sequential process to authorize the deployment of peacekeepers into Darfur first and then find a way to gain consent from Khartoum.

This was an unusual approach. This tension was illustrated by US Ambassador John Bolton during the negotiations:

> We said from the outset we did not contemplate that the UN peacekeeping force was going to fight its way into Darfur ... But we think it's important that the question of consent not hold up the operational steps that need to be taken to get this force deployed as rapidly as possible.
>
> (ST/AFP 2006b)

He expanded on this point in his memoirs. 'A traditional UN peacekeeping operation entailed the consent of all the parties ... that said, "consent by the Khartoum government" was a thinly veiled euphemism for never handing over to the UN,

---

[16] The other fundamental principles are impartiality and only using force in self-defence or to defend the mandate (DPKO 2008: 31).

which I knew [US President] Bush would simply not accept' (2007: 355). The US advocated this sequential process of 'resolution' then 'consent' to the UK and France (Author Interview 2011d). According to an official involved in the negotiations, the US was confident that consent would be forthcoming, while other Security Council members were not (Author Interview 2011a). States had good reasons to be sceptical of this approach; Security Council members received clear messages from the government of Sudan that not only did it not consent, but that it was openly hostile to UN peacekeeping.

These messages came directly from the president of Sudan on multiple occasions; for example, he stated in February 2006 that 'We are strongly opposed to any foreign intervention in Sudan and Darfur will be a graveyard for any foreign troops venturing to enter' (ST/AFP 2006a), and in August 2006 that Sudan would 'defeat any forces entering the country' (AP/ST 2006b).[17] This created an inherent tension in negotiations towards UN peacekeeping from the beginning. There was no consent from Khartoum—and indeed outright hostility—yet the political drivers towards UN peacekeeping remained, and the US drafters sought to find a way to meet disparate needs.

Unsurprisingly, there were real concerns within the Security Council over the lack of consent and how to handle this within a resolution. The issue of consent was not resolved before the resolution was tabled, and the resolution was facilitated by a text-based compromise rather than a substantive solution. China wanted standard language to be used: 'with the consent of the government of [country]' (see United Nations 2006b: 5).[18] However, the US did not want the word 'consent' to appear in the resolution, given the tensions outlined above. The UK brokered a deal to appease China whereby language on consent would be included if China allowed the UK to include language on the 'responsibility to protect', which was a key priority for the UK (US Permanent Mission to the UN 2006e). Three days prior to the final vote, Chinese UN negotiators were waiting on instructions from Beijing to see if it could accept this trade (US Permanent Mission to the UN 2006e). We can infer from the final version of resolution 1706 that Beijing accepted this trade, given that both consent and R2P appear in the final document.[19]

The language on consent was contentious for the US until the day before the vote. A US staff member reported that drafters resorted to using a thesaurus on the last day of drafting to find alternative words that were like 'consent'. Bolton wanted to use the term 'acquiesce' instead of 'consent' in an attempt, rhetorically at least, to

---

[17] While it is possible that Khartoum communicated a more flexible approach privately, a WikiLeaks cable confirms that Khartoum's hostility was communicated consistently in private meetings with both the DPKO and the African Union (US Permanent Mission to the UN 2006g).

[18] China's stance on the need for consent for resolution 1706 was in line with regional leaders (Teitt 2009)

[19] This inference is also consistent with a reciprocal trade of language made between China and the UK on R2P and other parts of the text in a presidential statement in April 2006 (US Permanent Mission to the UN 2006h).

lower the standard of consent required from Khartoum (Author Interview 2011e). The eventual text-based compromise was 'invites the consent of the Government of National Unity for this deployment', which did not resolve the underlying issue: Khartoum did not consent to peacekeeping in Darfur (United Nations 2006f: 3). As such, the political priority to transition to UN peacekeeping and the drive to 'do something' more effective towards civilian protection, combined with Khartoum's clear lack of consent, led to a text-based compromises which papered over the substantive issue of consent.

This text-based compromise created fundamental barriers to implementation. As DPKO outlines, a key component of credibility in peacekeeping is a 'clear and deliverable mandate' (2008: 38). The text-based compromise on consent created a resolution which was neither clear nor deliverable, with no guidance on how the resolution could be implemented given the lack of consent from Khartoum. Wood explains this fundamental challenge for Security Council resolutions: 'It is, of course, only possible to use clear language when the policy is clear' (1998: 82). Policy in this case was far from clear. In the words of one official involved in the negotiations, the language on consent allowed the Security Council to 'face both ways' (Author Interview 2011a). As US Ambassador John Bolton admitted in his memoirs, resolution 1706 'ducked the issue of what would happen if Khartoum declined to give its consent' (2007: 355). The Security Council prioritized passing a resolution over passing a resolution that could be implemented, and the use of text-based compromises facilitated this outcome. As such, it is an example of the external legitimation practice of 'doing something'.

Alongside the question of consent, the inclusion of language on R2P was highly contentious for Council members. In fact, a US State Department official who was involved in these negotiations went as far as to describe language on R2P as the most contentious part of resolution 1706 (Author Interview 2011e). The reason it was contentious was that it represented an evolution of language within Security Council resolutions. Once language has been agreed between Security Council members it becomes part of its standard repertoire and is likely to be used again in the future (for discussion see Gifkins 2016a, 2016b; OCHA 2014). R2P had been referenced in two Security Council resolutions prior to resolution 1706, however, these new negotiations were particularly contentious because the resolution applied R2P to a specific country and authorized international action in the form of a UN peacekeeping operation while referring to R2P.[20] The significance of this evolution in language draws from the precedent-based nature of Security Council decisions (Gifkins 2016a; Werner 2017). The subsequent failure of resolution 1706 weakened its precedent value, and it was not until 2011 that using R2P language in Security Council resolutions became commonplace (Gifkins 2016b:

---

[20] For further discussion of the negotiations between the UK, US, and China on the language of consent and R2P in resolution 1706 see Gifkins (2016a).

2–3). Resolution 1706 represents an early attempt by states that advocated R2P to normalize the concept within the Security Council.

Resolution 1706, mandating UN peacekeeping in Darfur, passed on 31 August 2006. It mandated the expansion of UNMIS from the south of Sudan (now South Sudan) into Darfur and allowed for interim support to AMIS. UNMIS in Darfur was mandated to support the implementation of the DPA and to prevent attacks against civilians within its capabilities (United Nations 2006f: 6). The resolution passed with the affirmative votes of twelve members of the Security Council, and was co-sponsored the same twelve members, as an extra level of legitimation from states in favour (United Nations 2006b: 2). China, Qatar, and Russia abstained from the vote for resolution 1706 (United Nations 2006b: 2). Funding pressures plus a desire to 'do something' led the drive towards the failed DPA and then failed resolution 1706. Khartoum was quick to reject the resolution and described it as 'unjustifiable hostility against Sudan' (AP/ST 2006a). This resolution exposed the lack of enforcement capacity the UN Security Council has when states refuse to comply, and has rightly been described as one of the 'worst Security Council resolutions ever' (Lynch 2010). This left the UN with an additional political problem as it exposed how easy it was for a country to refuse UN peacekeeping (Gowan 2008: 461–462).

## Towards Resolution 1769

### The Security Council at an impasse

The situation on the ground in Darfur deteriorated after resolution 1706 passed. In the aftermath of resolution 1706, Sudanese government-backed militia increased attacks on humanitarian staff as a way of resisting intervention by the international community (Flint and de Waal 2008: 197–198; Weissman 2007). Humanitarian actors were constrained and humanitarian staff decreased by about 18 per cent (Degomme and Guha-Sapir 2010: 298). This combined with an increase in banditry and fighting between rebel groups. In the year following resolution 1706, the number of internally displaced people in Darfur rose significantly (Degomme and Guha-Sapir 2010: 296). Although the increased instability did not result in an increased rate of violent deaths, it did increase overall mortality rates through preventable illness with fewer humanitarian staff on the ground (Degomme and Guha-Sapir 2010: 298). Reacting against resolution 1706, Khartoum restricted humanitarian actors, increasing the dire circumstances of internally displaced people.

The split curtailed action again in October, after Khartoum sent letters to traditional troop-contributing countries (TCCs) which stated 'in the absence of Sudan's consent to the deployment of UN troops, any volunteering to provide

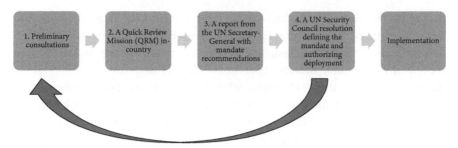

**Figure 8**  Authorizing Peacekeeping after the Failure of Resolution 1706

peace keeping [*sic*] troops to Darfur will be considered a hostile act, a prelude to an invasion of a member country of the UN' (US Permanent Mission to the UN 2006d). This letter was intended to prevent deployment by curtailing the availability of troops, and it worked. Jean-Marie Guéhenno said that, in response, member states had not been forthcoming with offers to contribute troops to the UN mission in Darfur (Lynch 2006). As DPKO relies on contributed troops to deploy UN peacekeeping, this was another way that Khartoum could resist troop deployment under resolution 1706. The president of the Security Council for October considered issuing a statement to condemn the letter, but members were again divided on how to respond (BBC 2006). With the Security Council at an impasse on how to respond to Khartoum's intransigence, it was unable to even unify around a statement against Sudan's hostility.

In theory, after the Security Council authorizes a peacekeeping operation, it will then source troops and equipment and deploy to the host country. Without Sudanese consent this was not possible. As illustrated in Figure 8, rather than move from authorization to implementation, as it usually would, the Security Council went from authorization back to the beginning of the cycle: preliminary consultations. Responding to the failure of resolution 1706, Security Council members worked closely with Khartoum to secure and maintain consent, and Khartoum had direct input into the drafting of subsequent resolution 1769. China increased its lobbying of Khartoum, as part of an external legitimation practice, and drafters refocused on the usual internal legitimation strategies of unanimity and regional support.

## Breaching the Impasse

With the Security Council at an impasse, an alternative strategy was needed to meet the political goal of replacing AMIS. The remarkable development is that the revised strategy came not from Security Council members but from the UN Secretariat itself, and the strategy was facilitated by Beijing encouraging Khartoum to agree to the plan. The strategic leadership in the Security Council of both the Secretariat and Beijing was unusual. The UN Secretariat has the capacity to shape decision-making either by normative developments or as a technical advisor to the Security Council. In this case, the office of the Secretary-General went beyond technical advice (on things like mandate) to design an alternative strategy to resolution 1706. As highlighted by the former Permanent Representative to Guatemala, when the Secretariat puts forward proposals that seem to have a high likelihood of success, Security Council members are likely to support the proposal giving the Secretariat greater capacity to influence decision-making (Rosenthal 2017: 109). Rather than simply offering technical advice on mandate size and mandate language, the Secretary-General created an alternative to resolution 1706 and instructed member states on how this plan could be achieved. Implementation of this plan was made possible by Beijing lobbying Khartoum.

In mid-November 2006 there were six weeks left of Kofi Annan's tenure as UN Secretary-General, which brought a renewed focus within his office on Darfur, R2P, and the legacy of his tenure (Author Interview 2011c; Hamilton 2011: 127). The Secretary-General's office created a Phased Approach Paper—which developed an alternative to resolution 1706—and told the P5 how important it would be that they present a unified front on this at an upcoming meeting in Addis Ababa (US Permanent Mission to the UN 2006c). The proposal had three stages: short-term assistance to AMIS, which was already being implemented, more substantial assistance to AMIS, including the deployment of UN personnel, and the deployment of an unprecedented hybrid AU–UN peacekeeping force (US Permanent Mission to the UN 2006c). This proposal became known as 'the Annan Plan'[21] due to the strong leadership on this issue from Kofi Annan himself. Annan was determined that an agreement would be reached at the meeting in Addis Ababa. The meeting was scheduled for two hours, but Annan planned to continue until an agreement was reached (US Permanent Mission to the UN 2006c). Indeed, Jean-Marie Guéhenno reported in his memoir that the meeting lasted for nine hours (2015: 200). At this meeting, the government of Sudan, represented by Foreign Minister Lam Akol, agreed in principle to the Annan Plan, with the caveat that issues on command and control and force size would have to be taken back to Khartoum for further discussion (Brookings Institute 2006). This represented a turning point in Khartoum's position. Annan has been singled out for praise on the leadership he showed at this meeting by US Envoy to Sudan Andrew Natsios (Brookings Institute 2006: 12 and 19). What is striking about this sequence of events is that the plan, leadership, and determination to breach the impasse came from the Secretary-General's office. P5 members agreed to support the plan, but it was the Secretariat, not states, which set the parameters for the negotiations.

---

[21] The name 'the Annan Plan' was also previously used in 2004 in relation to Cyprus.

The Annan Plan was crafted to appeal to the different interests of the actors involved. In this it was concessionary and appealed to the interests of Sudan and the AU more than the previous approach had. It had four key advantages for the varied actors involved. It allowed the AU to save face by being incorporated into the new operation (US Embassy Khartoum 2006). It was broad enough to let different parties see it differently—for example, to Western states it was a UN mission in all but name, and to China it was an AU mission (Author Interview 2011d). The plan allowed donors of AMIS to achieve their goal of sharing the cost of the mission through UN assessed contributions (Hamilton 2011: 79). And the hybrid was so much closer to Khartoum's preference for an African force that it even claimed that the hybrid was its idea (although the evidence does not support this) (Hamilton 2011: 128). Drawing from its position of legitimated authority, the Secretariat went beyond typical provision of advice to take leadership on designing and implementing a new strategy for peacekeeping in Darfur.

China's willingness to engage in bilateral diplomacy to encourage Khartoum's consent was pivotal in these negotiations. Host-state consent is a consistent priority for Chinese support of peacekeeping, however, actively encouraging consent itself represented an evolution of this priority. China helped to secure consent from the Sudanese government for the Annan Plan at the Addis Ababa meeting on 16 November 2006. There is clear recognition of China's role in gaining Sudanese consent in statements made by key insiders, such as Andrew Natsios the former US Envoy to Sudan, and by Stephen Morrison and Bates Gill from the Centre for Strategic and International Studies, all of whom testified before the US Committee for Foreign Affairs (US Committee on Foreign Affairs 2007: 5 and 21). Chinese Special Representative on Darfur Liu Guijin also stressed that bilateral diplomacy between Beijing and Khartoum was pivotal in securing Sudanese consent for the Annan Plan, and explained that the Chinese Foreign Ministry had endeavoured 'to take all possible opportunities and channels to work with the concerned parties, especially the Sudanese Government, in our own way and language' (Guijin 2007). Chinese Ambassador to the UN Wang Guangya was particularly singled out for his constructive role. Jean-Marie Guéhenno also has stressed the crucial role played by China in Addis Ababa (Brookings Institute 2006: 12). China played a pivotal role in securing Sudanese consent for the proposed hybrid peacekeeping operation, which can be understood as an external legitimation practice whereby Beijing sought legitimacy for itself in the eyes of key international audiences.

## Planning Hybrid AU/UN Peacekeeping

The importance of keeping Khartoum 'on board' was recognized during the planning stage for the hybrid, in a way that it was not during the earlier negotiations for UN peacekeeping. Although Khartoum had agreed to the Annan Plan 'in principle', it continued to delay and obstruct the process. The UN and AU worked with Khartoum to maintain consent. Newly appointed UN Secretary-General Ban Ki-moon and AU Commission Chair Alpha Oumar Konaré wrote to Khartoum in January 2007 requesting formal authorization for the second phase of the Annan Plan, while the first phase was being implemented. There were considerable delays in receiving a response from Khartoum (Ban 2007; US Permanent Mission to the UN 2007b). However, when the letter arrived from Khartoum, it reneged on the agreement it had made in November 2006 (Dagne 2007). The letter said that Khartoum would only accept peacekeeping if it were an AU mission, under AU command and control, financed by the UN, but subject to the Sudanese government's step-by-step approval for each component of the operation (US Embassy Khartoum 2007c). For it to be financed by the UN it needed to be a UN mission, so this was impractical. The UN and AU met again with Khartoum in April 2007 in Addis Ababa to address these concerns and to find a mutually agreeable solution (see United Nations 2007b: 11). The additional meetings were successful, and by the middle of April, Khartoum's position had changed and it had agreed to the second phase of the Annan Plan (United Nations 2007b: 11). After the failure of resolution 1706, Security Council members took a more conciliatory approach with Khartoum, recognizing that consent and unanimity were pivotal for deployment. While the previous negotiations took place largely in New York between Council members, these negotiations focused on working with the government of Sudan to seek and maintain its consent.

DPKO played an important role in the decision-making process of the Security Council by providing both information and recommendations to the Security Council. The limits of its influence were similar to those in the previous negotiations, however, in that its recommendations were followed, but there was political pressure to assume that the DPA was a viable peace agreement. The operational information that resolution 1706 was based on was months out of date. A Quick Review Mission (QRM) was conducted (in place of a TAM) jointly by the United Nations and African Union. This formed the basis of UN Secretary-General Ban Ki-moon's report, which was presented to the Council on 5 June 2007. As with the previous TAM, the DPKO drafters of the QRM report were under political pressure to proceed from the assumption that the DPA was viable, even though it was widely believed to be a failed agreement by this time (Holt, Taylor and Kelly 2009: 117–118). The QRM report recommended that the required hybrid force include 19,000 to 20,000 troops and 3,772 police officers (United Nations 2007b: 12). The Security Council followed this advice closely, as it had with the previous resolution, and the eventual mandate for the hybrid force was 19,555 military personnel and 3,772 police (United Nations 2007c: 3). The QRM reported that the security situation had deteriorated since the TAM was conducted, which was reflected in the recommendation of 13 per cent increase in the number of troops deployed (United Nations 2007b: 6–7). As in the previous negotiations, the Security Council followed DPKO's recommendations closely, which gives it considerable influence in shaping the mandate, however, DPKO was constrained politically by the political priorities of the Security Council to mandate peacekeeping, despite the evident issues in the DPA.

## A Different Approach to Drafting: Resolution 1769

Legitimation of the draft was reprioritized in the negotiations towards resolution 1769, particularly the usual internal legitimation practices of seeking unanimity and regional support. Consensus is particularly important for peacekeeping resolutions, which rely on legitimacy for implementation. In fact, the DPKO 'Capstone report' argues, in a section on the prerequisites for success in peacekeeping, that 'anything other than unanimous Security Council backing can be a serious handicap' (2008: 50). In response to the failure of resolution 1706—which was not unanimous and did not have consent from Khartoum—Security Council members reprioritized unity among themselves in drafting the next resolution and prioritized gaining and maintaining Khartoum's consent as paramount. In this sense, the failure of resolution 1706 reinvigorated the usual internal legitimation practice of unanimity. According to a UN source on Darfur, gaining consent for the hybrid peacekeeping operation was the single most significant breakthrough the UN made on Darfur (Author Interview 2011d). Drafters also prioritized regional support as an internal legitimation practice during the drafting process, whereby support from African states was courted and understood as important towards securing broader support, and quashing dissent. Unanimity between Council members and consent from Khartoum were recognized as paramount for implementation.

In negotiations towards the hybrid, consent was not taken as a given and Security Council members worked closely with Khartoum to gain and maintain consent. Following the QRM, further meetings were held in Addis Ababa between Khartoum, the UN Secretariat, and the AU to discuss the technical details of the plan and to alleviate Khartoum's concerns, and it was finally announced in June that the government of Sudan had consented to all three phases of the Annan Plan (UN News Centre 2007). After Khartoum's 'in principle' consent in November 2006 it had taken a further seven months of obstruction and delays before Khartoum before gave full consent. Even this delayed consent was only made possible by the conciliatory and collaborative approach taken by Security Council members during these negotiations. There were two contentious issues that drafters did not concede: acting under Chapter VII of the UN Charter[22] which allows the

---

[22] China often resists the use of Chapter VII (for discussion see Stähle 2008). China and Khartoum initially resisted the use of Chapter VII, however, Just before the resolution was finalized, under intense

Council to take legally binding measures, and unified command and control.[23] Security Council members had learnt from their previous experience and reprioritized consent and unanimity with the aim of not only passing a resolution but passing a resolution that could be implemented.

Drafters actively sought unanimity and consent throughout the negotiations towards resolution 1769 and were willing to make concessions to achieve it. Indeed, the Sudanese Ambassador to the UN said that he spent 'the whole of Sunday' at the British Permanent Mission going through the draft of resolution 1769 in depth before it was put to a vote, highlighting the high level of direct involvement Khartoum had in the Security Council's negotiations (Farley 2007). The UK, the US, and France wanted to include language which would put pressure on Khartoum, while China, Russia, Qatar, Indonesia, South Africa, and the Democratic Republic of the Congo wanted the draft to focus on positive steps Khartoum had taken (Security Council Report 2007: 5). The P3 conceded this, and the final text acknowledges Khartoum's positive steps. A full draft resolution from two weeks prior to the vote, available via US Permanent Mission to the UN. (2007c), offers unusual access to the full language that was under negotiation. Council unity and Khartoum's consent were prioritized, which is reflected in the conciliatory drafting process.

Four aspects of the resolution were conceded in the final two weeks of negotiations. First, the draft included a threat of 'further measures' in the event of non-compliance, suggesting sanctions, which did not appear in the final version. Second, a section was removed which attributed responsibility to 'rebel, Government and Janjaweed forces' for attacks on civilians and humanitarian workers. In the final version, attacks were condemned, but responsibility for violence was not discussed. Third, the draft specifically recalled the R2P agreement in paragraphs 138 and 139 of the World Summit outcome document. The UK wanted to refer to R2P in this resolution, as it had in resolution 1706, but both China and Sudan resisted this. Security Council unity was a higher priority for the UK than language on R2P, so it was removed (Author Interview 2012). Fourth, a reference to resolution 1706 was removed, as if erasing the previous failed decision. Resolutions typically list previous resolutions on the same topic, in order to demonstrate a history of legitimation (Werner 2017). However, references to failed resolution 1706

---

diplomatic pressure, China came onboard with the use of Chapter VII and, without China backing them, Khartoum agreed to accept a Chapter VII mandate (Ali 2007).

[23] The UK, China, Russia, and the US all wanted UN command and control and saw this as essential for UN funding (Reuters 2007; US Embassy Khartoum 2007d). The AU was reluctant to hand over full control to the UN, and Khartoum wanted it to remain an AU mission (US Permanent Mission to the UN 2007b). China actively engaged in bilateral diplomacy throughout the negotiations to encourage Khartoum to accept UN command and control (US Embassy Beijing 2007). A creative solution was needed to breech this impasse: resolution 1769 determined that there would be unity of command and control held by two officials with joint AU/UN appointment: Rodolphe Adada as Special Representative for Darfur and Martin Agwai as Force Commander (United Nations 2007c).

were opposed by China (US Embassy Paris 2007), South Africa (US Permanent Mission to the UN 2007c), the AU (US Permanent Mission to the UN 2007a), and Sudan (US Embassy Khartoum 2007b). To avoid mentioning resolution 1706, the preamble does not list the numbers for previous resolutions on Darfur. Resolution 1769 simply includes the text 'recalling all its previous resolutions and presidential statements concerning the situation in Sudan' (United Nations 2007c: 1). These conciliatory steps were taken on the understanding that Khartoum's consent and unity within the Council were paramount—not only for the resolution to pass but for it to be implemented.

Support from states in the region was also prioritized, as an internal legitimation practice, to strengthen the legitimacy of the decision and to help mitigate against dissent. Support was sought from African states and the African Union to bolster legitimacy for the hybrid. A US cable argued that gaining support from African Council members would be the key to overcoming resistance from other Security Council members (US Permanent Mission to the UN 2007d). Likewise, a senior AU official advised the US that the best way to proceed was for the AU to introduce a resolution calling on the United Nations to take over AMIS, arguing that AU support would 'bring China on board because it won't be holier than the Africans' (US Embassy Khartoum 2007a). Another US cable explained the P3 strategy for encouraging unanimous support:

> UKUN [the UK Permanent Mission to the UN] has said that if we can get AU Commission Chairperson Konaré to agree to our language, then we would have smooth sailing on the draft resolution not only among the P3 but among the wider Council Membership as well .... With Konaré on board, Ghana would likely align itself with our strong stance ... Should Ghana buy in, the other Council Members would be more likely to follow suit, which would make it all the more difficult for traditional obstructionists like China, Qatar, and South Africa to override their positions. South Africa, however, will require special massaging, but once on board, it would be that much more difficult for Sudan to challenge the Council's position.
>
> (US Permanent Mission to the UN 2007c)

These very candid reflections show the value that African support had in the eyes of Western drafters. Although Ambassador Khalilzad does not use the word 'legitimacy', what he is describing fits well with the way Coleman defines legitimacy. To paraphrase Coleman, support from African member states would generate a social status for the draft resolution because it was recognized as good or proper by African member states (2007: 20). Drafters thought that support from the AU, Ghana, or South Africa would place such a high level of legitimacy on the draft that it would be difficult for Sudan or other states to oppose the resolution. As such, internal legitimation practices were used as a negotiation tool. Drafters thought that if they could secure African support, they could use this to secure broader support or at least to mitigate dissent.

Resolution 1769 passed with the affirmative support of all fifteen Council members on 31 July 2007 (United Nations 2007c: 1–6). The resolution recognized that the government of Sudan had already consented to UNAMID, established a mandate, and requested that it take over from AMIS no later than 31 December 2007. The mandate included monitoring the arms embargo established under resolution 1556 and included steps to protect personnel and civilians. The resolution also outlined light and heavy support packages for AMIS in preparation for the hybrid operation. The statements made after the vote highlighted the cooperative approach that was taken with the government of Sudan in drafting this resolution (United Nations 2007a). Responding to the failure of resolution 1706, and the credibility gap it exposed, Security Council members reinvigorated their focus on consent and unanimity. Drafters worked closely with Khartoum to both obtain and maintain its consent throughout the negotiations. Drafters were also more willing to make concessions to secure broad support within the Security Council, such as removing language on R2P and on the threat of future sanctions.

Attacks on peacekeepers began almost immediately after deployment, and throughout, UNAMID faced exceptional obstruction from the government compared to other peacekeeping operations (Duursma 2021; Lynch 2014a, 2014b, 2014c; Read and Mac Ginty 2017). The expectations of what peacekeeping could achieve were also overinflated in relation to Darfur. Guéhenno explained: 'I knew that wat we could realistically try to achieve—if we were allowed to deploy in Darfur—would be only a limited show of force, to make it more complicated for those using violence to achieve their goals and to give a chance to politics' (2015: 195). Reflecting the challenging environment that UNAMID was deployed into, the UN High-level Independent Panel on Peace Operations (HIPPO) described it as 'a mere shadow of its original purpose' (United Nations 2015). As Lanz explains, 'Violence in Darfur did not end through a political settlement. Instead, the situation developed into a low-intensity, but highly protracted, conflict that is characterized by complex battlefield dynamics, fragmented rebel groups, the proliferation of tribal militias, widespread insecurity, and spurts of extreme violence' (Lanz 2020: 160). UNAMID, like AMIS before it, struggled with obstruction from the government and challenges related to the size and scale of the conflict in Darfur.

By 2012, the Security Council decided to decrease UNAMID's troop numbers in Darfur, even though violence was not substantially different than when it was authorized, and by 2017 there was increased pressure for an exit strategy (for discussion see Lanz 2020: 179–180). This was driven in part by a US goal to decrease the peacekeeping budget, for which it funds almost a third, and this campaign 'disproportionately impacted UNAMID, which was framed as the poster child of a mission that was no longer cost-effective' (Forti 2019: 7). This meant that the drawdown of UNAMID was not based on completion of its mandate but, like the establishment of the operation, it was based on competing pressures around funding. As Forti explains, 'the drawdown and transition were initiated before many of the core indicators of the mission's success were achieved' (2019: 8). Unlike the establishment of UNAMID, there was no longer a core civil society movement focused on Darfur to increase domestic pressure in the US and elsewhere (Lanz 2020: 179). An official who wrote a report for DPKO recommended withdrawal of UNAMID and argued that 'a large mission like UNAMID was not the right approach to armed conflict in Darfur. Because of the pressure from Save Darfur something large had to happen. But it was not based on a proper analysis of conflict causes in Darfur' (Lanz 2020: 180). Resolution 2429 in 2018 set out a plan to draw down UNAMID. with an intended exit in 2020 (United Nations 2018). In 2020, resolution 2559 was adopted, which ended the mandate of UNAMID at the end of the year (United Nations 2020). With considerably less fanfare than the establishment, in 2020 UNAMID quietly ceased to exist, with drawdown and liquidation finalized by 2022.

## Conclusion

There were two key turning points which facilitated the negotiations towards UNAMID—the Office of the Secretary-General designing the plan for hybrid peacekeeping and China lobbying Khartoum for consent. Both shifts were facilitated by legitimation practices. In the former, the Secretariat relied on its legitimated authority to develop, present, and encourage support for an alternative plan. The fact that the plan came from the Office of the Secretary-General (rather than a think tank or external lobby group) gave it the social status of legitimacy and made it more likely to be taken up by Security Council members. DPKO also had direct influence on the mandates set out in both resolutions 1706 and 1769 by way of recommendations set out in a report by the Secretary-General drafted by DPKO. In the latter, it was status-seeking that made Beijing receptive to lobbying from African states and Western states (Fung 2019). China was persuaded to actively seek consent from Khartoum, which represented an evolution in China's relationship with peacekeeping and consent. The legitimation of the Secretary-General's plan and China's external legitimation practices were critical in facilitating the creation of UNAMID.

While actions by China and the Office of the Secretary-General were the most significant turning points facilitating resolution 1769, there were also a range of additional 'moments' in the negotiations where external legitimation practices were evident and impactful. The first stage of negotiations was instigated by the desire to shift to a shared funding model via a transition to UN peacekeeping, but the methods pursued were shaped by a drive to 'do something', or be seen to be doing something, with substantial civil society activism on peacekeeping. It was evident prior to resolution 1706 that UN peacekeeping would not be viable in Darfur, with clear statements from Khartoum indicating its hostility to the idea. Peacekeeping is premised on consent, so drafters knew it could not be implemented before it passed, but proceeded anyway to 'do something'. Western drafters, particularly the US, sought the social status of legitimacy associated with taking action towards UN peacekeeping.

Actions in which internal legitimation practices were evident included drafters seeking support from African actors and a reinvigoration of the priority of unanimity. Drafters also sought support from African states—recognizing that support from the region would help facilitate other support and quash dissent. Cables from US diplomats at the time made frank statements making clear that they understood that support from African states was a means of legitimating the draft resolution. The final legitimation practice evident in these negotiations was the reinvigoration of the usual practice of unanimity. Responding to the failure of resolution 1706, drafters recognized that nothing less than a unanimous resolution would be sufficient and that gaining and maintain Khartoum's consent was an integral part of this. Unanimity and regional support were sought as internal legitimation practices towards the passage and implementation of resolution 1769.

Legitimation practices shaped the negotiations towards peacekeeping resolutions on Darfur, both in terms of external legitimation practices which facilitated decisions and internal legitimation practices during drafting. Decisions shaped by legitimation practices are not necessarily better in any normative sense or even implementable, however, as was also evident in relation to some of the negotiations towards sanctions. In drafting resolution 1706 as with the early sanctions resolutions, Security Council members found language that they could accept, rather than finding a substantive solution to the reality that Khartoum did not support UN peacekeeping in Darfur. In drafting the subsequent resolution that authorized UNAMID, drafters prioritized consent and unanimity, however, Khartoum continued to obstruct peacekeeping on the ground. Despite the challenges faced in implementation, legitimation practices are an unwritten part of Security Council negotiations and have a significant influence on both enabling and shaping decisions.

## Bibliography

Abbott, Kenneth W. and Duncan Snidal. 1998. 'Why States Act through Formal International Organizations'. *The Journal of Conflict Resolution* 42(1): 3–32.
African Union. 2004. *Agreement with the Sudanese Parties on the Modalities for the Establishment of the Ceasefire Commission and the Deployment of Observers in Darfur*. Addis Ababa, 28 May. Available at http://peacemaker.un.org/sites/
peacemaker.un.org/files/SD_040528_Agreement%20on%20the%20Modalities

%20for%20the%20Establishment%20of%20the%20Ceasefire%20Commission
%20in%20Darfur.pdf.

Ali, Wasil. 2007. 'Feature: Diplomatic Victory for US on Darfur Force Resolution'. *Sudan Tribune*, 31 July.

Allen, Susan Hannah and Amy T. Yuen. 2013. 'The Politics of Peacekeeping: UN Security Council Oversight across Peacekeeping Missions'. *International Studies Quarterly* 58(1): 621–632.

AP/ST. 2006a. 'Sudan Rejects UN Resolution on Darfur Peacekeepers'. *Sudan Tribune*, 1 September.

AP/ST. 2006b. *UN Force Would Risk Hezbollah-like Resistance—Sudan's Bashir*. Available at http://www.sudantribune.com/UN-force-would-risk-Hezbollah-like.

Author Interview. 2011a. *Author Interview with an Official Involved in the Negotiations*. 9 September, Copenhagen.

Author Interview. 2011b. *Author Interview with Former UN Under Secretary-General for Peacekeeping Affairs Jean-Marie Guéhenno*. 3 February, New York.

Author Interview. 2011c. *Author Interview with Member of the UN Department of Peacekeeping Operation*. 8 February, New York.

Author Interview. 2011d. *Author Interview with UN source on Darfur*. 25 August, London.

Author Interview. 2011e. *Author Interview with US State Department Official*. 15 February, New York.

Author Interview. 2012. *Author Interview with an Official Involved in the Negotiations*. 28 March, telephone interview.

Badescu, Cristina G. and Linnéa Bergholm. 2010. 'The African Union'. In *The International Politics of Mass Atrocities: The Case of Darfur*, eds. David R. Black and Paul D. Williams. London and New York: Routledge.

Ban, Ki-moon. 2007. *Informal Comments to the Media by the Secretary-General Ban Ki-moon on the Situation in Sudan and Other Matters*. New York. Available at http://webcast.un.org/ramgen/sc/so070215pm3.rm.

Barnett, Michael N. and Martha Finnemore. 2004. *Rules for the World: International Organizations in Global Politics*. Ithaca and London: Cornell University Press.

BBC. 2006. 'Sudan's Darfur Threat Condemned'. 1 October.

Bolton, John. 2007. *Surrender Is Not an Option: Defending America at the United Nations and Abroad*. New York: Threshold Editions.

Booth Walling, Carrie. 2013. *All Necessary Measures: The United Nations and Humanitarian Intervention*. Pennsylvania Studies in Human Rights. Philadelphia: University of Pennsylvania Press.

Brookings Institute. 2006. *Next Steps for Darfur*. Washington, D.C. Available at http://www.brookings.edu/~/media/events/2006/11/20africa/20061120.

Carlson, Allen. 2004. 'Helping to Keep the Peace (Albeit Reluctantly): China's Recent Stance on Sovereignty and Multilateral Intervention'. *Pacific Affairs* 77(1): 9–27.

Coleman, Katharina P. 2007. *International Organisations and Peace Enforcement: the politics of international legitimacy*. Cambridge: Cambridge University Press.

Dagne, Ted. 2007. *Sudan: The Crisis in Darfur and Status of the North-South Peace Agreement*. Congressional Research Service. Available at http://fpc.state.gov/documents/organization/84929.pdf.

de Waal, Alex. 2007. 'Darfur and the Failure of the Responsibility to Protect'. *International Affairs* 83(6): 1039–1054.

de Waal, Alex. 2007. 'Darfur's Elusive Peace'. In *War in Darfur and the Search for Peace*, ed. Alex de Waal: Global Equity Initiative, Harvard University and Justice Africa.

Degomme, Olivier and Debarati Guha-Sapir. 2010. 'Patterns of Mortality Rates in Darfur Conflict'. *Lancet* 375: 294–300.

Dijkstra, Hylke. 2014. 'Shadow Bureaucracies and the Unilateral Control of International Secretariats: Insights from UN Peacekeeping'. *Review of International Organisations* 10: 23–41.

DPKO. 2008. *United Nations Peacekeeping Operations: Principles and Guidelines*. United Nations. Available at https://peacekeeping.un.org/sites/default/files/capstone_eng_0.pdf.

Duursma, Allard. 2021. 'Pinioning the Peacekeepers: Sovereignty, Host-State Resistance against Peacekeeping Missions, and Violence against Civilians'. *International Studies Review* 23(3): 670–695.

Farley, Maggie. 2007. 'UN to Deploy Peacekeepers to Darfur'. *Los Angeles Times*, 1 August.

Flint, Julie and Alex de Waal. 2008. *Darfur: A New History of a Long War*. African Arguments. London and New York: Zed Books.

Forti, Daniel. 2019. *Navigating Crisis and Opportunity: The Peacekeeping Transition in Darfur*. International Peace Institute. Available at https://www.ipinst.org/wp-content/uploads/2019/12/1912_Transition-in-Darfur.pdf.

Fung, Courtney J. 2019. *China and Intervention at the UN Security Council: Reconciling Status*. Oxford: Oxford University Press.

Gifkins, Jess. 2016a. 'R2P in the UN Security Council: Darfur, Libya and Beyond'. *Cooperation and Conflict* 51(2): 148–165.

Gifkins, Jess. 2016b. *UN Security Council Resolutions and the Responsibility to Protect*. Asia-Pacific Centre for the Responsibility to Protect. Available at https://r2pasiapacific.org/filething/get/4949/%5BAP%20R2P%20Policy%20Brief%206.3%5D%20UN%20Security%20Council%20Resolutions%20and%20the%20Responsibility%20to%20Protect.pdf.

Gowan, Richard. 2008. 'The Strategic Context: Peacekeeping in Crisis, 2006–08'. *International Peacekeeping* 15(4): 453–469.

Guéhenno, Jean-Marie. 2015. *The Fog of Peace: A Memoir of International Peacekeeping in the 21st Century*. Washington, D.C.: The Brookings Institution.

Guijin, Liu. 2007. *A Positive Factor*. Available at http://www.bjreview.com.cn/qanda/txt/2007-07/16/content_69719.htm.

Haeri, Medina. 2008. 'Saving Darfur: Does Advocacy Help or Hinder Conflict Resolution?'. *Praxis: The Fletcher Journal of Human Security* XXIII: 33–46.

Hamilton, Rebecca. 2011. *Fighting for Darfur: Public Action and the Struggle to Stop Genocide*. New York: Palgrave Macmillan.

Hirono, Miwa and Marc Lanteigne. 2011. 'Introduction: China and UN Peacekeeping'. *International Peacekeeping* 18(3): 243–256.

Holslag, Jonathan. 2008. 'China's Diplomatic Manoeuvering on the Question of Darfur'. *Journal of Contemporary China* 17(54): 71–84.

Holt, Victoria, Glyn Taylor, and Max Kelly. 2009. *Protecting Civilians in the Context of UN Peacekeeping Operations: Successes, Setbacks and Remaining Challenges*. New York: United Nations.

Jacobsen, Katja Lindskov and Troels Gauslå Engell. 2018. 'Conflict Prevention as Pragmatic Response to a Twofold Crisis: Liberal Interventionism and Burundi'. *International Affairs* 94(2): 363–380.

Jago, Marianne. 2010. 'InterFET: An Account of Intervention with Consent in East Timor'. *International Peacekeeping* 17(3): 377–394.

Keating, Joshua E. 2010. 'Why Do Diplomats Still Send Cables?' *Foreign Policy*. Available at https://foreignpolicy.com/2010/11/30/why-do-diplomats-still-send-cables/.

Kent, Ann. 2002. 'China's International Socialization: The Role of International Organizations'. *Global Governance* 8(3): 343–364.

Lanteigne, Marc. 2011. 'A Change in Perspective: China's Engagement in the East Timor Peacekeeping Operations'. *International Peacekeeping* 18(3): 313–327.

Lanz, David. 2009. 'Commentary—Save Darfur: A Movement and Its Discontents'. *African Affairs* 108(433): 669–677.

Lanz, David. 2014. 'The Perils of Peacekeeping as a Tool of RtoP: The Case of Darfur'. In *Peacekeeping in Africa*, eds. Marco Wyss and Thierry Tardy. London: Routledge.

Lanz, David. 2020. *The Responsibility to Protect in Darfur: From Forgotten Conflict to Global Cause and Back*. Global Politics and the Responsibility to Protect. Abingdon and New York: Routledge.

Large, Daniel. 2008. 'China & the Contradictions of 'Non-interference' in Sudan'. *Review of African Political Economy* 115: 93–106.

Lee, Pak K., Gerald Chan, and Lai-Ha Chan. 2012. 'China in Darfur: Humanitarian Rule-Maker or Rule-Taker?'. Review of International Studies 38(2): 423–444.

Lynch, Colum. 2006. 'Sudan Escalates Stand Against UN Mission for Darfur'. *The Washington Post*, 6 October.

Lynch, Colum. 2010. *The 10 Worst UN Security Council Resolutions Ever*. Turtle Bay, Foreign Policy. Available at http://turtlebay.foreignpolicy.com/posts/2010/05/21/the_10_worst_un_security_council_resolutions_ever.

Lynch, Colum. 2014a. *A Mission That Was Set Up to Fail*. Foreign Policy. Available at http://foreignpolicy.com/2014/04/08/a-mission-that-was-set-up-to-fail/.

Lynch, Colum. 2014b. 'Now We Will Kill You'. Foreign Policy. Available at http://foreignpolicy.com/2014/04/08/now-we-will-kill-you/.

Lynch, Colum. 2014c. 'They Just Stood Watching'. Foreign Policy. Available at http://foreignpolicy.com/2014/04/07/they-just-stood-watching-2/.

MacKinnon, Michael G. 2010. 'The United Nations Security Council'. In *The International Politics of Mass Atrocities: The Case of Darfur*, eds. David R. Black and Paul D. Williams. London and New York: Routledge.

Mamdani, Mahmood. 2009. *Saviors and Survivors: Darfur, Politics and the War on Terror*. New York: Pantheon Books.

Morphet, Sally. 2000. 'China as a Permanent Member of the Security Council: October 1971–December1999'. *Security Dialogue* 31(2): 151–166.

Nathan, Laurie. 2007. 'The Making and Unmaking of the Darfur Peace Agreement'. In *War in Darfur and the Search for Peace*, ed. Alex de Waal. Global Equity Initiative, Harvard University and Justice Africa.

OCHA. 2014. *Aide Memoire for the Consideration of Issues Pertaining to the Protection of Civilians in Armed Conflict*. Available at https://docs.unocha.org/sites/dms/Documents/aide%20memoire%202014%20-%20English.pdf.

Orchard, Phil. 2010. 'Regime-Induced Displacement and Decision-Making within the United Nations Security Council: The Cases of Northern Iraq, Kosovo, and Darfur'. *Global Responsibility to Protect* 2(1–2): 101–126.

Read, Roisin and Roger Mac Ginty. 2017. 'The Temporal Dimension in Accounts of Violent Conflict: A Case Study from Darfur'. *Journal of Intervention and Statebuilding* 11(2): 147–165.

Reuters. 2007. 'Security Council to Back Funding for Darfur Mission'. Khartoum, 17 June. Available at http://www.sudantribune.com/spip.php?article22434.

Richardson, Courtney J. 2011. 'A Responsible Power? China and the UN Peacekeeping Regime'. *International Peacekeeping* 18(3): 286–297.

Rosenthal, Gert. 2017. *Inside the United Nations: Multilateral Diplomacy up Close.* Abingdon and New York: Routledge.

Security Council Report. 2007. *August 2007, Africa, Darfur/Sudan.* Available at http://www.securitycouncilreport.org/monthly-forecast/2007-08/lookup_c_glKWLeMTIsG_b_3041197.php.

ST/AFP. 2006a. 'Darfur Will be Foreign Troops' Graveyard—Bashir'. *Sudan Tribune*, 26 February.

ST/AFP. 2006b. *UN Security Council to Vote New Darfur Resolution Tomorrow.* Available at http://www.sudantribune.com/spip.php?article17337.

Stähle, Stefan. 2008. 'China's Shifting Attitude towards United Nations Peacekeeping Operations'. *The China Quarterly* 195(September): 631–655.

Teitt, Sarah. 2009. 'Assessing Polemics, Principles and Practices: China and the Responsibility to Protect'. *Global Responsibility to Protect* 1(2): 208–236.

Toga, Dawit. 2007. 'The African Union Mediation and the Abuja Peace Talks'. In *War in Darfur and the Search for Peace*, ed. Alex de Waal. Global Equity Initiative, Harvard University and Justice Africa.

UN News Centre. 2007. *Sudan Accepts Hybrid United Nations-African Union Peacekeeping Force in Darfur.* Available at http://www.un.org/apps/news/story. asp?NewsID=22881#.UIUvr2e-b0c.

United Nations. 2006a. *5439th Meeting United Nations Security Council.* New York, 16 May, S/PV.5439.

United Nations. 2006b. *5519th Meeting United Nations Security Council* New York, 31 August, S/PV.5519.

United Nations. 2006c. *Report of the Secretary-General on Darfur.* 28 July, S/2006/591.

United Nations. 2006d. *Resolution 1663.* 24 March, S/RES/1663.

United Nations. 2006e. *Resolution 1679.* New York, 16 May, S/RES/1679.

United Nations. 2006f. *Resolution 1706.* 31 August, S/RES/1706.

United Nations. 2006g. *Statement by the President of the Security Council.* 11 April, S/PRST/2006/16.

United Nations. 2007a. *5727th Meeting United Nations Security Council.* New York, 31 July, S/PV.5727.

United Nations. 2007b. *Report of the Secretary-General and the Chairperson of the African Union Commission on the Hybrid Operation in Darfur.* 5 June, S/2007/307/Rev.1.

United Nations. 2007c. *Resolution 1769.* 31 July, S/RES/1769.

United Nations. 2015. *Report of the High-level Independent Panel on Peace Operations on Uniting Our Strengths for Peace: Politics, Partnership and People.* Available at https://peacekeeping.un.org/en/report-of-independent-high-level-panel-peace-operations.

United Nations. 2018. *Security Council Resolution 2429.* 13 July, S/RES/2429.

United Nations. 2020. *Security Council Resolution 2559.* 22 December, S/RES/2559.

United Nations. 2022. UN Security Council Meetings & Outcomes Tables Dag Hammarskjöld Library. Available at https://research.un.org/en/docs/sc/quick/meetings/2022.

US Committee on Foreign Affairs. 2007. *The Escalating Crisis in Darfur: Are There Prospects for Peace?* Washington, D.C., One Hundred Tenth Congress, February 8, 110–116. Available at http://www.gpo.gov/fdsys/pkg/CHRG-110hhrg33109/pdf/CHRG-110hhrg33109.pdf.

US Embassy Beijing. 2007. *EAP DAS Tom Christensen with MFA North America DG Liu Jieyi*. Wikileaks. Available at http://wikileaks.org/cable/2007/07/07BEIJING4521.html.

US Embassy Khartoum. 2006. *Sudan/Darfur: Sudan Rejects UN PKO; UN Devises Darfur PKO Plan*. Wikileaks. Available at https://www.wikileaks.org/plusd/cables/06KHARTOUM1475_a.html.

US Embassy Khartoum. 2006g. Sudan/Darfur: Sudan Rejects UN PKO; UN Devises Darfur PKO Plan. Wikileaks. Available at https://www.wikileaks.org/plusd/cables/06KHARTOUM1475_a.html.

US Embassy Khartoum. 2007a. *AMIS Ready for an Honorable Exit*. Wikileaks. Available at http://wikileaks.org/cable/2007/05/07KHARTOUM780.html.

US Embassy Khartoum. 2007b. *Sudan Officials Differ on Darfur PKO, Political Process in Meetings with SE Natsios*. Wikileaks. Available at http://wikileaks.org/cable/2007/07/07KHARTOUM1090.html.

US Embassy Khartoum. 2007c. *Sudan Rejects UN Heavy Support Package*. Wikileaks. Available at http://wikileaks.org/cable/2007/03/07KHARTOUM392.html#.

US Embassy Khartoum. 2007d. CDA Hume Urges Change in Sudanese Position on Peace-Keeping in Darfur. Wikileaks. Available at https://wikileaks.org/plusd/cables/07KHARTOUM549_a.html.

US Embassy Ndjamena. 2006. *Chad/Darfur: AU-UN Joint Assessment Mission Briefs*. Wikileaks. Available at https://wikileaks.org/cable/2006/06/06NDJAMENA853.html.

US Embassy Paris. 2007. *Sudan—Special Envoy for Sudan Andrew Natsios' June 25 2007 Meeting with China Special Envoy for Sudan Liu Guijin*. Wikileaks. Available at http://wikileaks.org/cable/2007/07/07PARIS2912.html.

US Permanent Mission to the UN. 2006a. *Chinese look for Increased Cooperation as SC Mission to Sudan Departs*. Wikileaks. Available at https://wikileaks.org/plusd/cables/06USUNNEWYORK1149_a.html.

US Permanent Mission to the UN. 2006b. *Darfur Resolution Passes Unanimously with Some Chinese Help*. Wikileaks. Available at http://wikileaks.org/cable/2006/05/06USUNNEWYORK1017.html.

US Permanent Mission to the UN. 2006c. *P5 Meeting on UN SYG Sudan Proposal*. Wikileaks. Available at http://wikileaks.org/cable/2006/11/06USUNNEWYORK2157.html#.

US Permanent Mission to the UN. 2006d. *Sudan Mission in New York Threatens Potential Troop Contributors to Darfur Force*. Wikileaks. Available at http://wikileaks.org/cable/2006/10/06USUNNEWYORK1911.html#.

US Permanent Mission to the UN. 2006e. *UNSC/Darfur: Council Inching toward Adoption of Draft Resolution*. Wikileaks. Available at https://www.wikileaks.org/plusd/cables/06USUNNEWYORK1660_a.html.

US Permanent Mission to the UN. 2006f. *UNSC/Sudan: Proposed Way Forward in Securing Resolution for UN PKO in Darfur.* Wikileaks. Available at https://wikileaks. org/plusd/cables/06USUNNEWYORK1349_a.html.

US Permanent Mission to the UN. 2006h. UNSC/Sudan: PRST on Darfur Adopted Unanimously. Wikileaks. Available at https://www.wikileaks.org/plusd/cables/ 06USUNNEWYORK770_a.html.

US Permanent Mission to the UN. 2007a. *Khartoum Tightens Grip on AU and UNSC Over Hybrid.* Wikileaks. Available at http://wikileaks.org/cable/2007/07/ 07USUNNEWYORK579.html.

US Permanent Mission to the UN. 2007b. UNSC/Darfur: Still No Hybrid Clarity After AU Briefing. Wikileaks. Available at http://wikileaks.org/cable/2007/04/ 07USUNNEWYORK318.html.

US Permanent Mission to the UN. 2007c. *AU/UN Hybrid in Darfur: Narrowing Differences.* Wikileaks. Available at http://wikileaks.org/cable/2007/07/ 07USUNNEWYORK563.html#.

US Permanent Mission to the UN. 2007d. *Darfur Deployment: The View from New York.* Wikileaks. Available at http://wikileaks.org/cable/2007/03/ 07USUNNEWYORK193.html.

von Billerbeck, Sarah. 2019. '"Mirror, Mirror on the Wall": Self-Legitimation by International Organizations'. *International Studies Quarterly* 64(1): 207–219.

von Billerbeck, Sarah. 2020. 'No Action without Talk? UN Peacekeeping, Discourse, and Institutional Self-Legitimation'. *Review of International Studies* 46(4): 477–494.

Weissman, Fabrice. 2007. *Sudan Divided: The Challenge to Humanitarian Action.* New York: Médecins Sans Frontières. Available at http://www.doctorswithoutborders. org/news-stories/speechopen-letter/sudan-divided-challenge-humanitarian-action.

Wenqi, Zhu and Leng Xinyu. 2016. 'China in the Security Council'. In *The UN Security Council in the Twenty-First Century*, eds. Sebastian von Einsiedel, David M. Malone, and Bruno Stagno Ugarte. Boulder: Lynne Rienner.

Werner, Wouter. 2017. 'Recall It Again, Sam. Practices of Repetition in the Security Council'. *Nordic Journal of International Law* 86(2): 151–169.

Williams, Paul D. 2010. 'The United Kingdom'. In *The International Politics of Mass Atrocities: The Case of Darfur*, eds. Paul D. Williams and David R. Black. London and New York: Routledge.

Wood, Michael C. 1998. 'The Interpretation of Security Council resolutions'. In *Max Planck Yearbook of United Nations Law*, eds. Jochen A. Frowein and Rudiger Wolfrum. London: Kluwer Law International.

Wuthnow, Joel. 2010. 'China and the Processes of Cooperation in UN Security Council Deliberations'. *Chinese Journal of International Politics* 3(1): 55–77.

Xinhua. 2006. 'Chinese President Urges to Maintain Darfur Stability'. *Sudan Tribune.* Available at http://www.sudantribune.com/spip.php?article18468.

# Conclusion

To return to the motif used in the Introduction, the formal rules—as set out in the UN Charter and the Provisional Rules of Procedure—form a mere 'skeleton' for understanding the Security Council (Schia 2013: 140). These rules set out the basis of structure and stability but tell us little about how the Security Council operates on a day-to-day basis, and even less about how negotiations are conducted. While the formal rules are minimal, layered on top of the formal rules are informal practices which have evolved over decades and will continue to evolve. This book introduces two sets of informal practices that shape the process and outcome of Security Council negotiations. External legitimation practices—where an actor strives for legitimation from its key audiences which shapes its actions in the Security Council—and internal legitimation practices, which occur in negotiations towards enhancing the legitimacy of a decision. Analysing these two sets of legitimation practices—legitimation of the actor and legitimation of the decision—shows how the drive towards legitimacy in each of these areas were responsible for key turning points throughout negotiations on Darfur. In doing so, this analysis demonstrates the utility of shifting the focus of research on legitimation in international institutions, where the standard approach is to focus on the legitimacy of an institution or its decisions, towards the process of negotiation. This analysis demonstrates how the drive towards legitimacy shapes the decision-making process and the resulting decisions. In doing so, it demonstrates how internal and external legitimation practices both enabled decisions to be reached and shaped the resulting decisions.

Interest in the relationship between legitimacy and the Security Council has grown considerably in the past two decades; however, this literature has focused on the legitimacy of the Security Council as an institution or the legitimacy of its decisions (Binder and Heupel 2014; Chesterman 2006; Frederking and Patane 2017; Hurd 2002, 2007; Morris and Wheeler 2007; Ralph and Gallagher 2015; Welsh and Zaum 2013). Instead, this book analyses the way that legitimacy shapes negotiations and decisions. In doing so, it demonstrates the influence of the drive towards legitimation on Security Council negotiations, with Security Council members seeking legitimacy for themselves as actors and seeking legitimacy for its decisions. In the case of Darfur, these legitimation practices enabled deadlocks to be overcome, facilitated decisions, and shaped the resulting decisions. The argument here is that internal and external legitimation practices enable and shape

*Inside the UN Security Council.* Jess Gifkins, Oxford University Press. © Jess Gifkins (2023).
DOI: 10.1093/oso/9780192869029.003.0009

Security Council decisions, and understanding them is key to understanding the Council's decision-making.

Using internal and external legitimation practices as a lens shows why positions changed for key states during negotiations on Darfur, and why specific approaches were taken by drafters. As such, the focus of this book is on understanding Security Council decision-making, rather than critiquing this process, although the argument does point toward the informal and malleable nature of Council decision-making, which has implications for debates on reform. As highlighted in the Introduction, the book is not normative in focus; however, it does come from a position of wanting Security Council decisions to be more effective for the protection of civilians and for the responsibility to protect. My co-authors and I have developed normative arguments on Security Council negotiations elsewhere (Ralph and Gifkins 2017; Ralph, Gifkins, and Jarvis 2020). Internal and external legitimation practices are useful for understanding how Security Council negotiations are conducted, why positions can change during negotiations, and what processes are prioritized by drafters to secure maximum support.

This chapter summarizes the book and proceeds in the following three sections. First, a section on external legitimation practices—legitimation of the actor—showing how each of the key turning points in decisions on Darfur can be explained via actors changing position towards a stance that is recognized as legitimate by their key audiences. This is evidenced by examples across the issue areas analysed in the book: setting the agenda, sanctions, referral to the ICC, and peacekeeping. External legitimation practices also incentivize states to take action, in order to give the appearance that they are 'doing something', even when the actions they take cannot be implemented, as occurred in some decisions on sanctions and peacekeeping. The second section is on internal legitimation practices: legitimation of the decision. This section shows that it is much harder to evidence internal legitimation practices in relation to Darfur since it was an exceptional case rather than a routine case. However, there are examples of each of the internal legitimation practices—the drive towards unanimity, seeking regional support, and repetition of language—and situations where each of these practices were influential in overcoming a roadblock and facilitating a decision. The third section explains the relationship between internal and external legitimation practices and audiences. Internal and external legitimation practices were significant in enabling decisions to be reached on Darfur and shaped the resulting decisions throughout.

## External Legitimation Practices

The case study material in this book explains why states pursued policies they had previously refused to pursue or accepted policies they had previously refused to accept. These changes in policy are explained in this book as an outcome of a desire for legitimation by states from audiences that matter to them, whether that is domestic audiences or international peers. This argument stands in contrast to one of the key arguments on legitimation in relation to the Security Council, which is the argument put forward by Thompson that states channel their policies through the UN Security Council because they seek legitimation for their policies (Thompson 2006, 2010). In his argument, states do not change their policies, although they might accept minor alterations to them; powerful states simply use the Security Council as a tool for legitimating the strategy they were already planning to pursue. Thompson's argument can account for why states might choose to channel their plans through the Security Council for the legitimation benefit it brings, but it does not shed light on how states negotiate their policy inside the Security Council (beyond pursuing their interests) or account for changes in policies. The following section summarizes four key turning points in negotiations over Darfur in the Security Council, across each of the key tools used to address the conflict. In each of the following examples, states opposed an action before accepting or even championing that action, seeking legitimation from their key audiences. This shows that states are sensitive to legitimation of themselves as actors as part of international diplomacy and that this can change the actions that they pursue.

Each of the key turning points which facilitated Security Council decisions on Darfur were driven by concerns around legitimation for the actors involved. Across the four cases analysed in this book, there was initially resistance to action, either from Western states or China. This was overcome via a sensitivity to audience perceptions. First, it was Western states that resisted adding the situation in Darfur to the Security Council's agenda and ignored lobbying from OCHA on this. Mukesh Kapila, in his role as Humanitarian Coordinator for OCHA in Sudan, found that governments in the UK, the US, and elsewhere already had extensive intelligence showing the scale and nature of the violence in Darfur but did not want to add the situation to the agenda. After the nature of violence became public and generated public salience—particularly via media traction and via analogy to the Rwandan genocide—the UK and the US shifted positions and became champions for Darfur within the Security Council. It was when the information became public that there were legitimacy costs for Western states, with domestic publics as the relevant audiences, as Western states had extensive intelligence on the nature and scale of the violence and resisted Darfur as an agenda item regardless.

Second, on sanctions, the state most resistant to sanctions was China, although the US also resisted the application of targeted sanctions against specific individuals. It was China, along with Russia and Qatar, that ensured that the arms embargo only applied to Darfur, not the whole of Sudan, rendering it effectively unenforceable. However, it was a shift in China's position that enabled any individual targeted sanctions to be issued. China blocked the application of targeted sanctions when negotiations were conducted in the Sanctions Committee, which is a subsidiary body of the Security Council that is not open to public scrutiny. This suggests a relationship between external legitimation practices and the public or private nature of decisions. When decisions on targeted sanctions moved to the Security Council itself, Beijing then said it would be 'unthinkable' to block a decision against the wishes of fourteen other members (US Embassy Khartoum 2006). While we might not expect China, as an authoritarian regime, to be sensitive to domestic audience costs, in this instance it did demonstrate sensitivity to an international audience of its peers. By choosing to abstain, rather than veto, Beijing facilitated the application of targeted sanctions against four individuals in Sudan. The practical impacts of these targeted sanctions were close to nil (Dörfler 2019); however, it was a shift in China's position which enabled this decision to occur.

Third, the US was the chief opponent of the referral of Darfur to the International Criminal Court. By the time the Darfur referral was discussed, the US had spent years trying to carve out exemptions for US personnel, in the form of bilateral non-surrender agreements and by vetoing a resolution on peacekeeping in Bosnia and Herzegovina to obtain annual exemptions for personnel from non-signatory states. The US achieved some of what it wanted in the Darfur negotiations with a clause added to exempt personnel from states which are not a party to the Rome Statute from the ICC's jurisdiction. However, the US also maintained throughout these negotiations that it did not want to legitimize the ICC via its first referral from the Security Council. Until the day of the vote, it was not known whether the US would choose to veto the referral or allow the referral. In the end it proved too costly for the US to block a referral to the ICC of a case with high domestic salience and a case where the government had determined that a genocide had occurred and may be ongoing. It was sensitivity to legitimation from domestic public audiences and civil society groups, along with bilateral relationships with European states which advocated the referral, that led the US to abstain on this resolution.

Fourth, the initial attempt to transition from AU peacekeeping in Darfur to UN peacekeeping failed, however the eventual compromise in the form of hybrid UN/AU peacekeeping was implemented due to the UN Secretariat creating the plan and China bringing Sudan on board. The Secretariat, and in particular, the Office of the Secretary-General, leveraged its legitimated authority to develop, advocate, and strategize a plan and told Security Council members how to pitch it to Sudan. This plan gave states an alternative to the transition to UN peacekeeping, which had proved unimplementable. The political and practical drive towards UN peacekeeping remained, including donor frustration with AMIS and the need for broader civilian protection. With obstruction and reluctance, Khartoum consented to UNAMID, but it was bilateral diplomacy between Beijing and Khartoum that facilitated this consent. Western states and African states encouraged China to take this activist role—which it could have ignored—but it chose to shift position from protecting Khartoum in the Security Council to lobbying Khartoum outside of it. Here Beijing demonstrated 'status sensitivity' to use the words of Fung (2019: 64). Seeking the legitimation that comes from supporting the goals of peer states led to a change in Beijing's position, which in turn facilitated mandating and implementing UNAMID. What is remarkable in this case is that Beijing has historically been quite conservative on issues of consent in peacekeeping. Willingness to actively coerce consent from a host state represented an evolution in China's stance on peacekeeping.

This book analyses negotiations that were contentious, but where decisions were reached, rather than instances where it was not possible to reach a decision. The contentious nature of these decisions meant that there were sticking points, but these were necessarily overcome because a decision was reached (i.e., this is a study of decisions rather than non-decisions). To overcome these sticking points, a shift must have occurred. Key here, however, is the commonality of the shifts that enabled each of the above decisions to occur. Across four different issue areas and aspects of the Security Council's work, it was concerns around legitimation—to either domestic or international audiences—that facilitated a change in position for key actors in each of the negotiations and enabled the decision to pass. This shows that even at the 'high table' of international security, which is often understood solely as an arena for the national interest of its members, there is sensitivity to legitimation of states either by their domestic publics or by relevant international peers.

These examples show changes in state positions as an effect of legitimation concerns. These shifts enabled decisions to occur, but this does not necessarily mean that the decisions were 'better' in any normative sense by being shaped by legitimation, since this is not a normative evaluation, or even that the resulting decisions were implementable. The other component of external legitimation is the idea that states take action to 'do something', or to be seen to be 'doing something', whether the action is one that can be implemented or not. This was evident repeatedly in Security Council responses to Darfur, where a decision passed, but decision makers acknowledged that they knew at the time that it could not be implemented. Responding to legitimation pressures can lead states to want to 'look like' they are acting rather than necessarily having a strong interest in the actions being taken. Taking a decision in the Security Council gives the appearance of a strong action, and if the decision is subsequently not implemented that is likely to garner less attention. Prioritizing 'doing something' in the Security Council enables states to give the appearance of action and to signal to relevant audiences that they are acting.

Striving for legitimacy incentivizes 'doing something', or at least being seen to be doing something, even when these actions are not implementable. There are clear examples of this logic in the negotiations on sanctions and peacekeeping in relation to Darfur, where Western states, particularly the US, were determined to 'do something' to satisfy domestic audiences but had awareness at the time that the actions they were taking were unimplementable. John Danforth, for example, was US Ambassador to the UN when the first two resolutions which included the threat of sanctions passed (resolutions 1556 and 1564). Looking back on these negotiations, he said, 'there was no doubt in my mind that we were not going to impose sanctions because at least the Chinese would have vetoed the sanctions. I am 100% certain. There was no chance of getting sanctions' (Keane 2005).[1] Similarly, John Bolton was US Ambassador to the UN when the resolution passed on UN peacekeeping in Darfur (resolution 1706) and in his memoirs he described that, prior to the passage of resolution 1706, 'Khartoum, in many ways large and small, publicly and privately, was already making it clear it had no intention of accepting a UN peacekeeping force in Darfur, no matter what the [UN] DPA or the Security Council said' (Bolton 2007: 354). These examples highlight that key actors involved in the negotiations were aware that the decisions they drafted, championed, and authorized could not be implemented.

These two examples highlight that the goal in these instances was to 'do something', or to be seen to be doing something, rather than to authorize and implement the respective decisions, as people leading the negotiations recognized, correctly, that the decisions could not, and would not, be implemented. If decisions were passed with the forewarning that it could not be implemented, then the logical inference is that the goal was something other than implementation. Part of this will have been to increase pressure on Sudan, even if rhetorically, but a large part of this was about appealing to domestic audiences, especially in the US where the Darfur conflict had unusually high domestic salience (Hamilton 2011; Stedjan and Thomas-Jensen 2010). This is also connected to the external legitimation practice of consistent arguments. Once the US had established itself as a vocal advocate on Darfur, this incentivized the US to 'do something', or at least be seen to be doing something. As has been argued elsewhere, the US 'focused more on impression management for domestic audiences than devising a plan for dealing with Darfur' (Seymour 2014: 21). This 'impression management' is evidence of the government seeking the social status of legitimacy from a given audience, in this case, domestic audiences that were engaged with the situation in Darfur. By prioritizing 'doing something' over doing something that could be implemented, actors prioritized legitimation in the eyes of domestic audiences over assistance to people on the ground in Darfur.

This book demonstrates that external legitimation—states seeking legitimation from key audiences—led to shifts in state positions across a range of key issues on Darfur, from agenda-setting, to sanctions, and from the ICC referral to

---

[1] Resolution 1556 imposed some sanctions in the form of an arms embargo; however, here Danforth is referring to either targeted sanctions on individuals or broader sanctions against the oil industry which were discussed at the time.

peacekeeping. Key decisions were enabled by these changes in positions across the range of the Security Council's tools. These decisions were not necessarily good, or even implementable, and initial decisions on sanctions and peacekeeping could not be implemented. Seeking the social status of legitimacy led states to seek 'text-based compromises', where they found language they could accept, however this created situations where states had divergent interpretations of a resolution's meaning even at the time it passed. As Wood argued, 'language can only be clear when the policy is clear' (1998: 82). States found words that they could accept, on the threat of future measures under 'Article 41' and on 'inviting the consent' of Khartoum. Darfur had strong domestic salience, especially in the US, but was a lower foreign policy priority when compared to other foreign policy priorities in Sudan, which perhaps accounts for some degree of disconnect between strong rhetoric from Western states but limited action and unimplementable decisions (Williams 2010). In the US, the issue of Darfur generated one of the largest civil society movements ever, and has been described by former US Ambassador to the UN John Danforth, 'it was a very, very high priority, it was a remarkably high priority' domestically (Keane 2005). While the conflict in Darfur was, at best, third priority politically for the US within Sudan, behind its counterterrorism relationship with Khartoum and the North–South conflict (Stedjan and Thomas-Jensen 2010). This generated a dissonance whereby Western states—especially the US—were incentivized to 'do something' while also not having sufficient political investment in the situation to find workable actions.

## Internal Legitimation Practices

Among Security Council members, inside the Security Council, there are also internal legitimation practices evident in negotiations. Where external legitimation practices relate to actors changing position because they are seeking legitimation for themselves as actors, internal legitimation practices relate to drafters seeking enhanced legitimacy for individual Security Council decisions. The UN is replete with informal negotiation practices, and the Security Council is no exception (Wiseman 2015). There are a range of internal legitimation practices—including unanimity, regional relevance, and repetition of language—which are normal everyday occurrences in Security Council negotiations. These practices have both a normalized and strategic dimension. For example, unanimity is understood as the normal goal of Security Council negotiations, with 87 per cent of draft resolutions between 2000 and 2021 not only passing but passing unanimously.[2] Drafters also understand that there is a strategic value in achieving unanimity, and a cost in not achieving it (Rosenthal 2017; Schia 2017; von Einsiedel, Malone and

[2] Data compiled by the author from UN (2022).

Stagno Ugarte 2015). On regional support, there is an understanding that regional support is appropriate, however, seeking support from regionally relevant states can also be used strategically, as it was in the case of Darfur. Repetition of language is a normal everyday part of negotiations, as seen in OCHA's compendium of language on the protection of civilians in armed conflict; however, drafters also realize that it is easier to find consensus on language which has been 'previously agreed' so there can be a strategic value in using it (Dunne and Gifkins 2011; OCHA 2014, 2016; Werner 2017). For each of these internal legitimation practices there are both routinized and strategic dimensions in play. The Security Council's response to Darfur is a hard case for evidencing internal legitimation practices because it was one of the most contested cases of its era, and as such many dimensions of negotiations on Darfur were unique or exceptional rather than routine. There are, however, examples of each of these internal legitimation practices evident in the negotiations over Darfur.

First, unanimity is one of the hardest practices to evidence in the case of Darfur, because it was one of the most divisive cases discussed by the Security Council in the period 2004–2007, as demonstrated in Figures 1 and 2, however, signs of this practice are evident. Key resolutions on Darfur between 2004 and 2007 were all non-unanimous, except for resolutions 1679 and 1769, which were primarily focused on peacekeeping, although resolution 1679 also included a threat of future sanctions against individuals or groups who violated the DPA. The DPKO is clear that Security Council unanimity is important for implementing peacekeeping operations (DPKO 2008: 50). The resolution which authorized the TAM—1679—was unanimous, as is the normal practice. The subsequent resolution—1706—which authorized the failed transition to UN peacekeeping received abstentions from China, Qatar, and Russia (United Nations 2006: 2). Responding to this failure, however, drafters refocused on the usual practice of unanimity. It was clear from the different approach taken to negotiate resolution 1769—which authorized the hybrid peacekeeping operation—that unanimity was a key priority in these negotiations. Drafters were especially flexible when negotiating resolution 1769, as can be seen from the series of concessions they made, to ensure not only a unanimous resolution from the Security Council but also to secure and maintain consent from Khartoum. In response to the failure of resolution 1706, there was a reinvigorated focus on the usual practice of unanimous decisions, as it was recognized that it was important for both legitimation and implementation.

Second, language in Security Council resolutions typically draws heavily from 'agreed language', whereby it is easier to find agreement on phrases that members have agreed to previously. This tendency is so embedded that OCHA compiles 'agreed language' for drafters to draw from when negotiating new decisions (OCHA 2014, 2016). While repetition of language is a standard practice in Security Council negotiations, Darfur was an exceptional case in this regard as well. The contentious nature of negotiations on Darfur led to unusual formulations of 'text-based compromises', where states found language they could agree to even where there remained substantive disagreement on policy. These text-based compromises were most evident in negotiations on sanctions, consent in peacekeeping, and R2P. As in relation to unanimity, however, we can see traces of the practice of repeating language in negotiations on Darfur. One of the key enablers of the ICC resolution was the use of agreed language from a previous resolution on Liberia, resolution 1497. As part of the US's long-standing campaign to circumvent the authority of the ICC within the Security Council in 2003, the US insisted on language whereby individuals involved in peacekeeping who are from a state that is not a party to the Rome Statute are not subject to the jurisdiction of the ICC for crimes committed in the host state. This language was then repeated in 2005 in the referral of Darfur to the ICC, which led France to distance itself from the resolution by withdrawing its sponsorship of the draft and led Brazil to abstain from the resolution. Despite some states' concerns around this language, it is difficult to refuse language once it has been agreed in another context. As such, while negotiations on Darfur were exceptional due to their contentious nature, there are still examples where the practice of repetition of language proved crucial to passing a decision.

Third, support from states in the region under discussion is particularly prized as an internal legitimation practice. In the early years of the Darfur conflict, the African Union led the strongest response to the crisis in the form of leading peace negotiations and providing AU peacekeeping (Badescu and Bergholm 2010). Within the Security Council, there were examples of drafters prioritizing support from African states because of the legitimacy that their support brings to a decision. When drafting resolution 1679, the US described co-chairing negotiations with the Congolese delegation as a means of building broader support for the draft (US Permanent Mission to the UN 2006). Likewise, when drafting resolution 1769, there were examples of drafters seeking support from African states and the AU as a strategy of gaining broader support from other Security Council members and limiting dissent (US Embassy Khartoum 2007; US Permanent Mission to the UN 2007a, 2007b). For example, a US cable argued that AU support would 'bring China on board because it won't be holier than the Africans' (US Embassy Khartoum 2007). Another cable stressed that African support was critical for securing the unanimous support from Security Council members beyond Africa (US Permanent Mission to the UN 2007a). These examples highlight the strategic legitimation value that drafters see in obtaining support from regionally relevant states.

While these three internal legitimation practices had less impact on the resulting decisions than the two external legitimation practices above, it is still clear that they helped to facilitate decisions and shaped the parameters of the decisions. Support from African states was courted strategically to garner broader support, previously agreed language was used to overcome a roadblock in the negotiations, and unanimity was reprioritized after the failed transfer to UN peacekeeping. Darfur is a difficult case with which to evidence internal legitimation practices because it was one of the most contentious items on the Security Council's agenda at the time. While there is evidence that these internal legitimation practices shaped and facilitated decisions at times, they likely have greater impact on other cases which are more routine and less unique. The Security Council's negotiations on Darfur show that even in a highly contentious case, these routine internal legitimation practices are still influential in shaping and facilitating outcomes.

## Internal and External Legitimation Practices

Internal and external legitimation practices both help to enable decisions to be reached and affect the specifics of the resulting decisions. As such, they are an area that has not been given sufficient attention in literature on Security Council negotiations. Analysing Security Council negotiations on Darfur via the lens of external legitimation practices has shown that there is a sensitivity for states on how key audiences will respond to actions taken inside the Security Council. The specific audiences that are valued vary among states, with Western states showing sensitivity to domestic audiences in this case, and China showing sensitivity to its international peers. While the audiences vary, sensitivity to audience perceptions was key in enabling decisions to be made across the areas of agenda-setting, sanctions, referral to the ICC, and peacekeeping. Analysing decisions through the lens of internal legitimation practices show that drafters are sensitive to how a decision will be perceived and recognize that both broad and targeted support is necessary to enhance the legitimacy of a decision. There is clear recognition in the literature that the Security Council has limited capacity to enforce compliance with decisions and relies on perceptions of legitimacy to encourage compliance with decisions (Hurd 2007; Johnstone 2008; Zaum 2013). Internal legitimation practices are one of the techniques that drafters use to enhance the perceived legitimacy of a decision, which in turn increases the chances of compliance with the decision. These two types of legitimation practices enable decisions to be made and determine the scope of what is possible in a resolution.

As noted above, legitimation practices do not necessarily mean that the resulting decisions are normatively better, or even implementable. Darfur was a prominent topic in the Security Council between 2004 and 2007, and decisions were regularly made to apply new tools to the situation. The text-based compromises on threatened sanctions and on the transition to UN peacekeeping could not be implemented. However, even when decisions could be implemented, the impact on the ground has often been minimal. Targeted sanctions, in the form of a travel ban and an asset freeze, were issued against only four individuals and have not been enforced (Dörfler 2019). The implementation of UNAMID peace-keeping faced an extreme version of 'pinioning', where obstruction from the host government seriously curtailed the activities of peacekeepers (Duursma 2021). Peacekeepers in Darfur came under attack from the early stages of deployment, and there was self-censorship from UNAMID for fear of being further restricted by Khartoum (Lynch 2014a, 2014b, 2014c). The referral to the ICC has similarly made little impact and has yet to hold anyone accountable for crimes commit-ted in Darfur (Kersten 2016). Reaching any decisions on Darfur took extensive negotiations in the Security Council and often compromises in the form of find-ing language members could agree to. Legitimation practices did not necessarily make decisions any easier to implement, and often made them harder to imple-ment. Even where decisions were reached, there was limited impact on the ground. Legitimation practices are a feature of multilateral negotiations and compromise, and can result in decisions that cannot be implemented.

Legitimation practices could be a useful lens for a range of other multilateral settings or domestic settings. In multilateralism, it is a useful tool for assessing the ways in which legitimation shapes negotiations, both in terms of the legitima-tion of actors involved in negotiations as well as the legitimation of the decisions produced in that forum. It could also have purchase at the domestic level, for research on domestic governance and the ways in which legitimation of actors and decisions affects the process and outcome of negotiations. Other possible areas of research in relation to the UN Security Council include applying legitimation practices to other case studies. It would be interesting to analyse to what extent external legitimation practices are evident in cases with less domestic salience or whether there is a spectrum of external legitimation practices across case stud-ies related to the level of domestic interest. Another possible direction for future research in relation to the Security Council is further exploration of the effects of the drive towards unanimity. The move towards consensus decision-making in the Security Council has been described as one of 'two remarkable trends of the post-Cold War era', the other being increased use of Chapter VII, and the impli-cations of the pull towards unanimity across case studies has not yet been fully explored (von Einsiedel, Malone, and Stagno Ugarte 2015). Chapter 2 set out cur-rent patterns of unanimity and examples to demonstrate a sense of compliance pull towards affirmative voting, and there is scope to explore this further with a meta-study of cases.

This book has demonstrated the utility in connecting research on practice the-ory with research on legitimacy and legitimation. An area that would be interesting to explore further is the relationship between practices, legitimation, and critique. One possibility is that 'actors often accept nonoptimal (i.e., inefficient and/or inequitable) procedures and practices as the way things are, while new practices that may be more efficient but are not embedded in the status quo often disappear' (Johnson, Dowd, and Ridgeway 2006: 68). The legitimation practices identified in this book contributed towards outcomes that, at times, could not be implemented, and where they could be implemented, had serious limitations on the ground. This raises questions for further research about the potential for critique within practice theory, the capacity for informal change of practices, and the relationship between competence and practice (for discussion see Adler-Nissen and Pouliot 2014; Hopf 2018; Ralph and Gifkins 2017; Schindller and Wille 2019). The connections between practice theory and legitimation are multifaceted and raise lots of potential avenues for future research.

This research demonstrates the role that external and internal legitimation practices have in both facilitating and shaping Security Council decisions. External legitimation practices facilitated decisions after periods of inaction. This occurred when states shifted their positions due to sensitivity to relevant audiences and the social status they would gain by acting. External legitimation practices are also evident when states make decisions, even if not implementable, to look like they were 'doing something'. These external legitimation practices facilitated reaching decisions but did not necessarily result in decisions which were implementable, with states finding 'text-based compromises' to paper over substantive disagreement. Internal legitimation practices shaped the resulting decisions. Chapter 2 set out reasons to expect that internal legitimation practices are common and widespread in day-to-day negotiations in the Security Council. However, Darfur was a highly contentious case, and therefore negotiations were more exceptional than they were routine. Nevertheless, there are examples of internal legitimation practices throughout the negotiations on Darfur. Across the case studies, there were examples where states reprioritized unanimous decision-making, sought support from regionally relevant states to facilitate broader support, and used previously agreed language to circumvent disagreement. Internal legitimation practices have both strategic and normalized dimensions whereby drafters seek unanimity, regional support, and use agreed language because these are understood as the routinized and correct ways to conduct negotiations, and because they have strategic value in enhancing the legitimacy of a decision, especially when it is a contentious decision. Internal and external legitimation practices enable and shape decisions at the highest levels of international politics.

The focus on internal and external legitimation practices has meant that this book arrived at different conclusions to much of the existing literature on Darfur negotiations. As set out in the introductory chapter, literature on Security Council negotiations in relation to Darfur and beyond primarily describes negotiations as an outcome of national interests, especially material interests, of states. This book presents an alternative view, whereby shifts in state positions are accounted for by states seeking the social status of legitimacy (external legitimation practices) and smaller turning points are explained by drafting practices intended to achieve maximum legitimacy for the decision (internal legitimation practices). In doing so, it also presents an alternative narrative of the roles that P5 members took in relation to Darfur. There is a tendency in literature on Darfur negotiations to frame China as obstructionist and Western states as advocates. This book unsettles that binary understanding, however, with instances where Western states were obstructionist (on agenda-setting, on which individuals targeted sanctions were applied to, and on the ICC referral), while China facilitated decisions in some instances (both unanimous resolutions towards peacekeeping were facilitated by China). An analysis of legitimation practices shows how factors other than material interest were influential in shaping state actions and areas in which the binary narrative of Western advocates and Chinese obstruction does not fit.

Beyond the literature on Darfur, this analysis represents a shift in institutional analysis on legitimation. Where other literature on legitimation and the Security Council focuses on legitimacy of the Council as an institution or the legitimacy of its decisions, this book demonstrates the utility of an alternative approach: how legitimation operates within negotiations. The drive for legitimation of states as actors and the drive for maximizing the legitimacy of an individual decision are shown in this analysis to have impacts on both how decisions are reached and the content of those decisions. As such, legitimation practices within negotiations are an underresearched area, worthy of further study in relation to other case studies as well. Where earlier literature on legitimation of the Security Council as an institution can explain why states seek to associate themselves—and their favoured policies—with the Council, and literature on legitimation of the Council's decisions can explain why states comply with its decisions, this research explains how particular decisions were reached and the impact of legitimation practices on the resulting decisions. In doing so, it shows that much of the negotiations on Darfur can be explained via an analysis of legitimation practices and that social status for actors and for decisions is sufficiently influential to determine the outcome of many negotiations. While states demonstrated certain 'red lines' over substantial sanctions in the case of China and the possibility of prosecution of citizens in the case of the US, beyond these few constraints, much of the decision-making was determined via legitimation practices. Further research on internal and external legitimation practices could explore this in relation to other cases in the Security Council or other institutional settings.

This research also has implication for the perennial debates over reform of the Security Council. These debates generally focus on membership and voting reform, and while there is broad consensus that formal reform of the UN Charter is necessary, there is insufficient consensus on any given model of reform (Hurd 2008; Nadin 2016; Weiss and Young 2005). Given this, some of the most interesting debates over Security Council reform are those that focus on informal change, particularly on the Security Council's working methods (Niemetz 2015; Weiss 2003; Wenaweser 2016). This book demonstrates that the process and outcome of negotiations is strongly shaped by informal practices; reform or change in this area does not require an amendment to the UN Charter. It shows that there is some capacity for media and civil society to influence Security Council negotiations via domestic actions, although this influence may have different impacts from what was intended (Hamilton 2011; Lanz 2009, 2020). While this book does not advocate specific policy prescriptions for reform, it does demonstrate that internal and external legitimation practices are key in shaping decisions, which opens scope for informal change. Legitimation practices are central to the production of Security Council decisions, and any change in this area circumvents negotiations over composition and voting in the Council.

# Bibliography

Adler-Nissen, Rebecca and Vincent Pouliot. 2014. 'Power in Practice: Negotiating the International Intervention in Libya'. *European Journal of International Relations* 20(4): 889–911.

Badescu, Cristina G. and Linnéa Bergholm. 2010. 'The African Union'. In *The International Politics of Mass Atrocities: The Case of Darfur*, eds. David R. Black and Paul D. Williams. London and New York: Routledge.

Binder, Martin and Monika Heupel. 2014. 'The Legitimacy of the UN Security Council: Evidence from Recent General Assembly Debates'. *International Studies Quarterly* 59(2): 238–250.

Bolton, John. 2007. *Surrender Is Not an Option: Defending America at the United Nations and Abroad*. New York: Threshold Editions.

Chesterman, Simon. 2006. 'Reforming the United Nations: Legitimacy, Effectiveness and Power after Iraq'. *Singapore Year Book of International Law* 10: 1–28.

Dörfler, Thomas. 2019. *Security Council Sanctions Governance: The Power and Limits of Rules*. Routledge Research on the United Nations. Abingdon and New York: Routledge.

DPKO. 2008. *United Nations Peacekeeping Operations: Principles and Guidelines*. United Nations. Available at https://peacekeeping.un.org/sites/default/files/capstone_eng_0.pdf.

Dunne, Tim and Jess Gifkins. 2011. 'Libya and the State of Intervention'. *Australian Journal of International Affairs* 65(5): 515–529.

Duursma, Allard. 2021. 'Pinioning the Peacekeepers: Sovereignty, Host-State Resistance against Peacekeeping Missions, and Violence against Civilians'. *International Studies Review* 23(3): 670–695.

Frederking, Brian and Christopher Patane. 2017. 'Legitimacy and the UN Security Council Agenda'. *PS: Political Science and Politics* 50(2): 347–353.

Fung, Courtney J. 2019. *China and Intervention at the UN Security Council: Reconciling Status*. Oxford: Oxford University Press.

Hamilton, Rebecca. 2011. *Fighting for Darfur: Public Action and the Struggle to Stop Genocide*. New York: Palgrave Macmillan.

Hopf, Ted. 2018. 'Change in International Practices'. *European Journal of International Relations* 24(3): 687–711.

Hurd, Ian. 2002. 'Legitimacy, Power, and the Symbolic Life of the UN Security Council'. *Global Governance* 8(1): 35–51.

Hurd, Ian. 2007. *After Anarchy: Legitimacy and Power in the United Nations Security Council*. Princeton: Princeton University Press.

Hurd, Ian. 2008. 'Myths of Membership: The Politics of UN Security Council Reform'. *Global Governance* 14: 199–217.

Johnson, Cathryn, Timothy J. Dowd and Cecilia L. Ridgeway. 2006. 'Legitimacy as a Social Process'. *Annual Review of Sociology* 32(1): 53–78.

Johnstone, Ian. 2008. 'Legislation and Adjudication in the UN Security Council: Bringing Down the Deliberative Deficit'. *The American Journal of International Law* 102: 275–308.

Keane, Fergal. 2005. *John Danforth Interview Transcript*. BBC. Available at http://news.bbc.co.uk/2/hi/programmes/panorama/4647211.stm.

Kersten, Mark. 2016. *Justice in Conflict: The Effects of the International Criminal Court's Interventions on Ending Wars and Building Peace*. Oxford: Oxford University Press.

Lanz, David. 2009. 'Commentary—Save Darfur: A Movement and Its Discontents'. *African Affairs* 108(433): 669–677.

Lanz, David. 2020. *The Responsibility to Protect in Darfur: From Forgotten Conflict to Global Cause and Back*. Global Politics and the Responsibility to Protect. Abingdon and New York: Routledge.

Lynch, Colum. 2014a. *A Mission that Was Set Up to Fail*. Foreign Policy. Available at http://foreignpolicy.com/2014/04/08/a-mission-that-was-set-up-to-fail/.

Lynch, Colum. 2014b. *'Now We Will Kill You'*. Foreign Policy. Available at http://foreignpolicy.com/2014/04/08/now-we-will-kill-you/.

Lynch, Colum. 2014c. *'They Just Stood Watching'*. Foreign Policy. Available at http://foreignpolicy.com/2014/04/07/they-just-stood-watching-2/.

Morris, Justin and Nicholas J. Wheeler. 2007. 'The Security Council's Crisis of Legitimacy and the Use of Force'. *International Politics* 44: 214–231.

Nadin, Peter. 2016. *UN Security Council Reform*. Global Institution Series. Abingdon and New York: Routledge.

Niemetz, Martin Daniel. 2015. *Reforming UN Decision-Making Procedures: Promoting a Deliberative System for Global Peace and Security*. Routledge Research on the United Nations. Abingdon and New York: Routledge.

OCHA. 2014. *Aide Memoire for the Consideration of Issues Pertaining to the Protection of Civilians in Armed Conflict*. Available at https://docs.unocha.org/sites/dms/Documents/aide%20memoire%202014%20-%20English.pdf.

OCHA. 2016. *Aide Memoire for the Consideration of Issues Pertaining to the Protection of Civilians in Armed Conflict*. Available at https://www.unocha.org/sites/unocha/files/Aide%20Memoire%202016%20II_0.pdf.

Ralph, Jason and Adrian Gallagher. 2015. 'Legitimacy Faultlines in International Society: The Responsibility to Protect and Prosecute after Libya'. *Review of International Studies* 41(3): 553–573.

Ralph, Jason and Jess Gifkins. 2017. 'The Purpose of United Nations Security Council Practice: Contesting Competence Claims in the Normative Context Created by the Responsibility to Protect'. *European Journal of International Relations* 23(3): 630–653.

Ralph, Jason, Jess Gifkins and Samuel Jarvis. 2020. 'The UK's Special Responsibilities at the United Nations: Diplomatic Practice in Normative Context'. *The British Journal of Politics & International Relations* 22(2): 164–181.

Rosenthal, Gert. 2017. *Inside the United Nations: Multilateral Diplomacy up Close*. Abingdon and New York: Routledge.

Schia, Niels Nagelhus. 2013. 'Being Part of the Parade—"Going Native" in the United Nations Security Council'. *Political and Legal Anthropology Review* 36(1): 138–156.

Schia, Niels Nagelhus. 2017. 'Horseshoe and Catwalk: Power, Complexity, and Consensus-Making in the United Nations Security Council'. In *Palaces of Hope: The Anthropology of Global Organizations*, eds. Ronald Niezen and Marie Sapignoli. Cambridge: Cambridge University Press.

Schindller, Sebastian and Tobias Wille. 2019. 'How Can We Criticize International Practices?'. *International Studies Quarterly* 63(4): 1014–1024.

Seymour, Lee J.M. 2014. 'Let's Bullshit! Arguing, Bargaining and Dissembling Over Darfur'. *European Journal of International Relations* 20(3): 1–25.

Stedjan, Scott and Colin Thomas-Jensen. 2010. 'The United States'. In *The International Politics of Mass Atrocities: The Case of Darfur* eds. Paul D. Williams and David R. Black. London and New York: Routledge.

Thompson, Alexander. 2006. 'Coercion through IOs: The Security Council and the Logic of Information Transmission'. *International Organization* 60(1): 1–34.

Thompson, Alexander. 2010. *Channels of Power: The UN Security Council and US Statecraft in Iraq*. New York: Cornell University Press.

United Nations. 2006. *5519th Meeting United Nations Security Council* New York, 31 August, S/PV.5519.

United Nations. 2022. *UN Security Council Meetings & Outcomes Tables* Dag Hammarskjöld Library. Available at https://research.un.org/en/docs/sc/quick/meetings/2022.

US Embassy Khartoum. 2006. *Sudan/China: UNSC Vote on Darfur Shows Limits of Friendship*. Wikileaks. Available at https://wikileaks.org/plusd/cables/06KHARTOUM996_a.html.

US Embassy Khartoum. 2007. *AMIS Ready for an Honorable Exit*. Wikileaks. Available at http://wikileaks.org/cable/2007/05/07KHARTOUM780.html.

US Permanent Mission to the UN. 2006. *UNSC/Sudan: Proposed Way Forward in Securing Resolution for UN PKO in Darfur*. Wikileaks. Available at https://wikileaks.org/plusd/cables/06USUNNEWYORK1349_a.html.

US Permanent Mission to the UN. 2007a. *AU/UN Hybrid in Darfur: Narrowing Differences*. Wikileaks. Available at http://wikileaks.org/cable/2007/07/07USUNNEWYORK563.html#.

US Permanent Mission to the UN. 2007b. *Darfur Deployment: The View from New York*. Wikileaks. Available at http://wikileaks.org/cable/2007/03/07USUNNEWYORK193.html.

von Einsiedel, Sebastian, David M. Malone, and Bruno Stagno Ugarte. 2015. *The UN Security Council in an Age of Great Power Rivalry*. United Nations University. Available at https://i.unu.edu/media/cpr.unu.edu/attachment/1569/WP04_UNSCAgeofPowerRivalry.pdf.

Weiss, Thomas G. 2003. 'The Illusion of UN Security Council Reform'. *The Washington Quarterly* 26(4): 147–161.

Weiss, Thomas G. and Karen E. Young. 2005. 'Compromise and Credibility: Security Council Reform?'. *Security Dialogue* 36(2): 131–154.

Welsh, Jennifer and Dominik Zaum. 2013. 'Legitimation and the UN Security Council'. In *Legitimating International Organizations*, ed. Dominik Zaum. Oxford: Oxford University Press.

Wenaweser, Christian. 2016. 'Working Methods: The Ugly Duckling of Security Council Reform'. In *The UN Security Council in the Twenty-First Century*, eds. Sebastian von Einsiedel, David M. Malone, and Bruno Stagno Ugarte. Boulder: Lynne Rienner.

Werner, Wouter. 2017. 'Recall It Again, Sam. Practices of Repetition in the Security Council'. *Nordic Journal of International Law* 86(2): 151–169.

Williams, Paul D. 2010. 'The United Kingdom'. In *The International Politics of Mass Atrocities: The Case of Darfur*, eds. Paul D. Williams and David R. Black. London and New York: Routledge.

Wiseman, Geoffrey. 2015. 'Diplomatic Practices at the United Nations'. *Cooperation and Conflict* 50(3): 316–333.

Wood, Michael C. 1998. 'The interpretation of Security Council resolutions'. In *Max Planck Yearbook of United Nations Law*, eds. Jochen A. Frowein and Rudiger Wolfrum. London: Kluwer Law International.

Zaum, Dominik. 2013. 'International Organizations, Legitimacy, and Legitimation'. In *Legitimating International Organizations*, ed. Dominik Zaum. Oxford: Oxford University Press.

# Index

*For the benefit of digital users, indexed terms that span two pages (e.g., 52–53) may, on occasion, appear on only one of those pages.*